Driving Women

Fiction and Automobile Culture in Twentieth-Century America

DEBORAH CLARKE

The Johns Hopkins University Press

Baltimore

© 2007 The Johns Hopkins University Press
All rights reserved. Published 2007
Printed in the United States of America on acid-free paper

2   4   6   8   9   7   5   3   1

The Johns Hopkins University Press
2715 North Charles Street
Baltimore, Maryland 21218-4363
www.press.jhu.edu

Library of Congress Cataloging-in-Publication Data
Clarke, Deborah, 1956–
Driving women : fiction and automobile culture in twentieth-century
America / Deborah Clarke.
p.   cm.
Includes bibliographical references and index.
ISBN-13: 978-0-8018-8550-1 (hardcover : alk. paper)
ISBN-13: 978-0-8018-8617-1 (pbk. : alk. paper)
ISBN-10: 0-8018-8550-7 (hardcover : alk. paper)
ISBN-10: 0-8018-8617-1 (pbk. : alk. paper)
1. Automobiles—Social aspects—United States. 2. Women automobile
drivers—United States. 3. Automobiles in literature. I. Title.
HE5623.C53 2007
813'.509355—dc22   2006022815

A catalog record for this book is available from the British Library.

In memory of my father, Bruce Clarke,
who passed on to me his love of cars

and for Phil

Many people have contributed to this book, so many, in fact, that I cannot hope to name them all. Numerous colleagues left material in my mailbox pertaining to the project, and the book is much richer for it. Even when I elected not to include something, the enthusiasm and interest reassured me about the worth of the project. People really are fascinated by car culture and gender.

Students in various classes also offered valuable responses and information. A few shared extremely useful resources, particularly Melanie Monroe. Lauren Carnali helped to track down material on topics from homelessness to the United Farm Workers. I owe special thanks to my research assistant of several years, Kristin Jacobson, whose help was invaluable in finding material that I didn't even know existed, checking out resources, and searching magazines for car advertisements. Her generosity and good spirits helped keep me going.

Kate Brogan, Susan Harris, Carla Mulford, Marcy North, and Robin Schulze read drafts of chapters, offering important criticism and much valued support. Janet Lyon, Mark Morrisson, and Robin Schulze helped me to define my arguments and meet my writing goals. And to Sue Schweik, who read the entire manuscript with painstaking care, offering suggestions that were always just right, I owe more than I can ever articulate or repay.

Cathleen R. Latendresse and Linda Skolarus at the Research Center of the Henry Ford Museum were extremely useful in helping me to track down archival material that related to women; their interest and aid were vital at an early stage of the project. With their support I discovered a wealth of material at The Henry Ford. In addition, I'd like to thank librarians at the Automobile History Collection of the Detroit Public Library, the Walter Reuther Archives at Wayne State University, and Yale University's Beinecke Library for valuable assistance.

I'm grateful to Michael Lonegro of the Johns Hopkins University Press for his strong support of the project and appreciate all his efforts to make the process easier and the book better. Penn State University provided sabbatical support as well as travel money for visits to Detroit, and I'm particularly thankful to the Penn State Institute for the Arts and Humanities for a Resident Fellowship that procured for me additional writing time. An early draft of Chapter One appeared under the same title in *Arizona Quarterly* 59.4 (Winter 2003): 103–33; part of Chapter Four appeared as "Domesticating the Car: Women's Road Trips" in *Studies in American Fiction* 32.1 (Spring 2004): 101–28; and a very small portion of Chapter Two is drawn from "William Faulkner and Henry Ford: Cars, Men, Bodies, and History as Bunk," *Faulkner and His Contemporaries,* ed. Ann Abadie & Donald Kartiganer (Jackson: UP of Mississippi, 2004), 93–112.

Finally, I thank my husband, Phil Bolda, not only for his patience, his understanding, and his love but also for all the Web searches that turned up valuable information regarding the auto industry and other sundry topics. I know it wasn't easy living with a book-obsessed spouse, but he managed it with grace and good humor, and, most important, knew enough not to ask how the book was coming. Phil, you have made my life—and my work—better.

Driving Women

Writing and Automobility

Few objects epitomize American identity more than the automobile. Although it was not born in the U.S.A., we have claimed the car as our own. As Cynthia Dettelbach has noted, "In America, the automobile shapes—and haunts—the imagination" (120). The car therefore shapes and haunts American literature. It also shapes and haunts our material lives. According to the Alliance of Automobile Manufacturers, the auto industry and its feeder industries account for roughly 5 percent of all private sector jobs in the country and more than 3.7 percent of its gross domestic product. By 2001 nearly 92 percent of all U.S. households owned at least one vehicle (Hu & Reuscher 32). Not many people would challenge the claim that Americans have become increasingly dependent on their cars, not just for getting around but also for making a statement about who they are. Nothing announces—and defines—one's presence more than an automobile. To quote Stephen Bayley, "For some people, owning a new car is the nearest they will ever get to perfection in an otherwise flawed and soiled life" (4). Both material necessity and imaginative icon, the automobile plays a critical role in defining American identity and serves as a crucial yet remarkably overlooked element within twentieth-century American fiction.

Cars offer convenience, comfort, and, above all, power. But in our culture they seem to offer such qualities primarily to men. Women still report being patronized by car sales personnel and intimidated by auto mechanics. Popular myth associates cars with masculinity, and automobile advertising continues to link the car to the female body, promising men control over speed and women. The shadow the automobile casts is a masculine one. As John Steinbeck muses in *Cannery Row,* "Someone should write an erudite essay on the moral, physical, and aesthetic effect of the Model T Ford on the American nation. Two generations of Americans knew more about the Ford coil than the clitoris, about the planetary system of gears than the solar system of stars" (69). For Steinbeck "generations" seem to comprise men, whose primary interest is automotive; women, one hopes, may know more about the clitoris. But Steinbeck has a point in observing not only that knowledge of cars (and cars themselves) takes precedence over knowledge of women but that both women and cars are objects of cultural scrutiny. The association of women and cars is an integral element of car culture. Historian Beth Bailey cites a 1951 article in *Time* magazine on attitudes toward premarital and extramarital sex, quoting a man who remarked: "After marriage, some guy taking my wife would be like taking my car and putting on a few extra miles. It might improve through use, but I like to drive my own." The "equation of wife and car," Bailey notes, "passed without comment in the resulting story" (70).

Given this assumption of women's position in car culture—under male control—it is no surprise that despite the fact that women now purchase more than 50 percent of the nation's cars, one encounters few women race car drivers or mechanics, and even car saleswomen remain in a distinct minority. Ford Motor Company did not have a woman on its board until 1976 (Butler 15); by 2005 only seven of fifty-two top corporate officials were women, giving Ford the highest percentage in the U.S. auto industry (Schneider). According to the consulting group Catalyst, in 2005 "just over 11 percent of top executives in auto-related companies were women, compared with more than 22 percent in publishing, nearly 16 percent in pharmaceuticals, 14 percent in railroads and 15 percent in mail and freight delivery" (Schneider). For an industry with such visibility in American society, women remain remarkably invisible within it.

Women may lack visibility in the auto industry, but they do not lack presence in automobile culture, a presence far more compelling than simply being an extension of the auto body. In response to a comment by an irate male listener about why they always seem to side with female callers, Click and Clack—aka Tom and Ray Magliozzi, hosts of National Public Radio's massively popular *Car Talk*—declare, "We think it is incumbent upon us to tell the American public that in our experience fixing cars, women are better at describing problems, better at answering questions about symptoms, and, in general, have less of their egos tied up in pretending that they know everything" ("Dear"). Women are perfectly capable of negotiating the automotive world. Further, if women are less invested in cars, having less of their egos tied up in them, that may explain why they are such acute observers of cars. This power of observation pays off when it comes to writing literature: not only are women capable of understanding the ways their cars run; they are also capable of articulating what their cars mean. Cars are very visible in women's fiction once a reader begins to look for them.

For the past hundred years American women have been writing about cars. This is hardly surprising, given the extent to which the car defines and shapes American culture. To be American is to have a stake in automobility. But to be an American woman is to be both car and driver, both object and subject of automotive culture. Thus, while the few existing studies of the automobile in American literature tend to focus primarily, if not exclusively, on male writers, female writers face a far more daunting task in detailing the influence of the car on women's lives and women's literature. This is not to suggest that it's easy if you are a man—as if car equals penis equals power is, for example, a valid equation. Indeed, writers such as Harry Crews, William Faulkner, F. Scott Fitzgerald, Jack Kerouac, Sinclair Lewis, and John Steinbeck have each explored multiple and fascinating elements of automobile culture. There's truth, after all, to the proposition that most men care about cars. But women writers must confront the assumption that they have no place in automobile culture, that women know little—and care less—about cars. Such thinking is very far from the truth. As automotive journalist Lesley Hazleton puts it, "Here is a machine that produces power on demand—real, immediate, physical power, tangible in all the senses. You could almost call it a machine specifically designed to enhance women's growing exploration of the possibilities for freedom and control of their own lives" (*Confessions* 43). In particular, women's literature stands as a testament to the intense interest women take in car culture: the power it conveys, the inequities it reveals, and the dangers it entails.

Driving Women contends that women participate in automobile culture in intricate and complex ways and that American women's fiction proffers a register of those responses over the course of the twentieth century and beyond. In providing access to the public sphere—to work, to escape—the car transformed women's lives as profoundly as suffrage. Thus, women's relation to cars serves as an important site in which issues critical to twentieth-century American culture—technology, mobility, gender, domesticity, and agency—are articulated. Throughout the century popular discourses surrounding the automobile govern cultural perceptions of the role of the car in determining identity, place, and gender. Women writers, responding to these cultural formations, help to shape that meaning, writing women firmly into an automotive culture from which much popular sentiment has sought to exclude them. By exploring the presentation of the car in women's fiction, one can measure the extent of women's intervention in the age of the automobile. The intersection of women and cars provides a vantage point from which to trace the shifts and rifts in the notion of woman's place, agency, and identity across the century.

The automobile, note Peter Marsh and Peter Collett in *Driving Passion: The Psychology of the Car,* "satisfies not only our practical needs but the need to declare ourselves socially and individually" (5). That declaration of individuality is crucial. Automobile historian James Flink contends that "individualism—defined in terms of privatism, freedom of choice, and the opportunity to extend one's control over his physical and social environment—was one of the most important American core values that automobility promised to preserve and enhance in a changing urban-industrial society. Mobility was another" (*Car* 38). By accentuating the potential for individuality, the car allows women a position from which to construct individual identity, exercise individual agency, and chart a course as acknowledged individuals in American culture. The power of the machine grants a form of mobility hitherto unimagined for women; even the bicycle demanded considerable physical strength and kept women reasonably close to home. The car, however, extends and erases the boundaries of the home, further obliterating what remained of the notion of separate spheres. As historian Virginia Scharff argues, women in cars, "employing the multiple possibilities of the automobile, gave new meaning to the notion of 'woman's place'" (164).

When the car becomes a woman's place, all bets are off. This is certainly not to say that cars have offered women unconditional freedom and liberation; women have often paid a considerable price for the benefits of automobility. Between chauffeuring, commuting, dealing with accidents, and financing maintenance and repairs, many women are just as enslaved to their automobiles as they ever were to their homes. The car, indeed, poses very real dangers to women: to life and limb, including the threat of sexual assault, as detailed, for example, in Kathy Dobie's memoir, *The Only Girl in the Car.* Joyce Carol Oates, in "Lover," reminds us, however, that cars can also serve as weapons for women; her protagonist uses her car to instigate a traffic accident in the hopes of killing the man who has broken off their affair. She buys a car specifically for this purpose and realizes that "the Saab was free of human weakness; its exquisite machinery was not programmed to contain any attachment to existence, any terror of annihilation" (168). The unnamed protagonist matches the car's "freedom from human weakness," illustrating that women with cars can be just as deadly as men with cars. The power of automobility comes in all shapes and sizes—and in a range of ethical codes.

Almost as soon as they appeared, cars opened up a new world, both materially and imaginatively, allowing women a wider range of possibility in their everyday lives and their literature. While men also gained tremendously from the rise of the automobile, their traditional access to the public sphere, exercise of individual power and agency, and relation to mobility and technology had never been as constrained as women's; for them it was more a shift of degree than of kind. Thus, the role of the automobile resonates particularly strongly in women's fiction. It is not a simple extension of individual power and control; rather, it functions as a contested site in which the very notion of femininity is challenged and ultimately reformulated. As cars become more and more enmeshed into our lives, their influence becomes increasingly complicated, destabilizing conventional ideology surrounding the home, gender, race, ethnicity, and "Americanness."

Investigating women's relation to the car generates new ways of thinking about American culture and American configurations of gender. Given the enormous influence of the automobile in shaping twentieth-century American culture, this is a topic that begs for consideration. Wondering why there are so few studies of the meaning of the car, Hazleton suggests that "unless we can understand the power that cars and driving have over our lives—what keeps us dependent, what pulls us to cars, what excites us about them—any critique of them remains on the purely intellectual level, divorced from the reality of our lives and our roads. The more unaware we are of how cars affect us, the less we will be able to confront and control the damage they do to our minds, and to our planet" (*Confessions* 118–19). Cars, in other words, must be considered on the material as well as the imaginative level. We must consider them as vehicles as well as symbols, as "real" things that affect real people.   *symbolic*

Studying the automobile in women's fiction may seem to force cars precisely back into the realm of the imaginary, the literary, and thus to disengage them from the so-called real world. Literature, however, both responds to and shapes material reality. Focusing on literary discourse may necessarily narrow the scope of this study, but it also provides a lens from which to view the developing relationship between women and cars across a century. Writers, like the rest of us, are very much a part of car culture, and a close look at the texts makes clear the extent to which these women have understood the underlying assumptions regarding gender and the automobile. Further, in setting the literature against a larger cultural context—advertising, journalism, girls' books, TV, film, music, cultural and feminist theory—this book attempts to map out the role of fiction in furthering our understanding of how American culture operates and what various forces work together to produce the ideology surrounding the automobile and its status as a gendered icon. Each chapter begins with a discussion of how the automobile inflects a particular angle of American culture— race, maternity, or citizenship, for example—then considers how such concerns have been articulated in the literature. Writers do not simply respond to culture; they help to shape it through support, revision, and resistance. Situating women writers' use of the car against a cultural backdrop opens up new avenues into women's literature and women's lives.

Exploring these avenues demands that one investigate a wide range of novels written over the course of the twentieth century. The car, of course, does not function as a universal symbol any more than it provides all things to all people. Numerous factors—including race, ethnicity, class, geography, age, family status, sexual orientation, and economic position—intersect to determine the effect of the car on an individual. As the range and number of texts included here makes clear, the assumption that women have scant interest in automobiles and what they mean is proved patently false. On the contrary, women writers make clear their understanding of the ways that women's lives are imbricated with automotive culture. Considering women's fiction from this angle also illustrates that women's literature is fully engaged with issues of technology, industry, and mobility—and the ways that such concerns intersect with identity, maternity, and the home. It helps us to continue the feminist work that challenges claims of women's exclusion from the interests of the public sphere and all that it encompasses: economics, business, citizenship, and politics. The novel, so often seen as a female genre, becomes a vehicle for fiction of mobility rather than domesticity, of technology rather than sentimentality.

Narrative fiction provides a particularly useful perspective for the study of automobile culture because of the breadth of experience it can represent. Cars, after all, shape our lives in numerous ways: materially, economically, imaginatively. The novel allows for a richly nuanced presentation of the range of meanings associated with the car. This is not to suggest that other genres and media, especially film, cannot offer equal complexity, but, as a scholar of the novel, I have chosen to concentrate my attention on fiction, with a few brief forays into memoir and travel narrative. Ellen Garvey, investigating women and bicycles, notes that because "the fears women's bicycling raised were social, fiction, with its articulation of social relationships, was better adapted than medical or other articles to taking the sting out of those fears by fictionally reconfiguring the relationships the bicycle seemed to be changing and by assigning new meanings to those changes" (84). Fiction exploring the car goes much further, challenging attempts to domesticate the automobile rather than assuaging the anxieties attendant upon it. The books discussed here do not, by any means, constitute an exhaustive list, but they do offer exceptionally compelling and intricate insights into the meaning and influence of the automobile. The primary subject of this book is contemporary fiction, though the first two chapters explore the origins of automobile culture and its gendering in the early twentieth century. The patterns and assumptions put in place in the first two decades of the appearance of the automobile in the United States persist even today. This study touches upon the work of roughly twenty-five female authors: Dorothy Allison, Julia Alvarez, Kay Boyle, Joan Didion, Louise Erdrich, Jessie Fauset, Leslie Feinberg, Cristina Garcia, Zora Neale Hurston, Cynthia Kadohata, Michelle Kennedy, Barbara Kingsolver, Rose Wilder Lane, Erika Lopez, Bobbie Ann Mason, Toni Morrison, Joyce Carol Oates, Flannery O'Connor, Ann Petry, Marge Piercy, Danzy Senna, Leslie Marmon Silko, Mona Simpson, Jane Smiley, Gertrude Stein, Helena Viramontes, and Edith Wharton.

Because it is often useful to play women's presentation of the car against men's, there is some discussion of male writers as well: Russell Banks, Harry Crews, Don DeLillo, William Faulkner, F. Scott Fitzgerald, Richard Ford, William Kennedy, Jack

Kerouac, Sinclair Lewis, John Steinbeck, Jean Toomer, and Richard Wright, though only Faulkner receives extended treatment. Overall, women seem less concerned about the car's relation to sexuality, a topic of far greater interest to men. Interestingly, a focus on sexuality tends to emerge most powerfully in texts in which the motorcycle, rather than the car, functions as the primary vehicle. Even more strikingly, the motorcycle is often tied to lesbian identity, as in the case of Feinberg and Lopez. So, while cars remain the primary subject, motorcycles provide a fascinating, if brief, glance at an alternative form of mobility.

Driving Women begins not so much with the history of the automobile, a well-documented subject, but with the ways in which the automobile has both reflected and constructed particular ideas about gender in early-twentieth-century American culture. While there is an abundance of secondary material about cars, little of it mentions women explicitly. Primary sources, however, such as popular journalism, advertising, and children's books, along with more "literary" texts, tell a very different story. Both excited and frightened by the rapidity with which the automobile transformed American culture, contemporary observers displaced the fears and hopes of technology onto women, looking to them to domesticate the age of automobility. As the century progressed, the car became increasingly incorporated into the literature; thus, the last four chapters focus primarily on contemporary fiction set against a range of cultural discourses such as advertising, TV, and various studies of automobile culture. Just as the car itself has been altered to adapt to changing times—pollution concerns, safety concerns, lifestyle issues, taste—so the car's meaning and influence must be mapped against particular historical contexts. Yet there is one constant: the automobile remains a contested site for American women's lives and American women's fiction.

The range of material, time, and authors addressed here demands a range of methodological approaches. Consequently, I draw on feminist theory, automobile history, technology studies, mobility studies, and cultural studies to help guide my exploration of the ways in which automobile culture and women's fiction mutually inform each other. Quite simply, cars abound in so much of our culture that to cover every aspect of automobility would demand a multivolume study. Thus, other artistic outlets such as music, visual art, film, and television receive short shrift, and more attention is devoted to advertising and the auto industry itself. Given the extent to which ads become ingrained in our heads, they seem to have the widest and strongest impact in shaping our awareness of cars and car culture. We can probably all recite some of the most memorable ad slogans: "Baseball, hotdogs, apple pie, and Chevrolet"; "Wouldn't you really rather have a Buick?"; "Built Ford tough"; "It's not your father's Oldsmobile"; "Oh what a feeling—Toyota." As Jennifer Wicke has observed,

"Advertisements are cultural messages in a bottle" (17). Exploring the traces of such messages in literature proves very profitable. Ads, according to Roland Marchand, "contributed to the shaping of a 'community of discourse,' an integrative common language shared by an otherwise diverse audience" (xx). The "common language" of automobile advertising reveals many of the underlying gender codes that women writers often end up debunking in their fiction.

In addition, various statistical findings on car ownership and women drivers, as well as material from the industry on how to appeal to women, provide a fascinating glimpse into the immediate influence of the car on women's lives and the ways in which the industry attempts to shape that influence as its agents consider the age-old question of what women want. (In this respect they appear to share the ignorance of the knight in Chaucer's "Wife of Bath's Tale," but in his case he finally listens to a woman and discovers that women desire "maistrie" above all else; the auto industry might be well advised to read the tale.) Finally, as the Chaucer example makes clear, this book relies upon the power of reading. Few items illuminate and challenge our culture as thoroughly as a literary text—which makes literature an excellent venue from which to explore the product that has done so much to define twentieth-century America. These novels offer valuable insight into the innumerable moments of intersection between women's literature and automotive culture.

The first chapter interrogates the symbolism of the car in early-twentieth-century American culture, examining how the development of the automobile industry reflects and creates many of the major concerns facing the country and how women became a crucial point of articulation around which these tensions were enacted. In Chapter Two I investigate modernism, gender, and race, looking at the role of the automobile in challenging white male dominance by giving women and nonwhite men control over technology. Chapter Three deals with the ways that automotive technology transforms maternity, arguing that the slippage between woman and machine creates not just a cyborg maternity but also an automotive maternity: material and mobile, natural and mechanical. The next chapter looks at issues of mobility and travel, tracing how the car reshapes domesticity and female agency. Drawing on feminist mobility theory, I posit that women's road trips transform domesticity from being situated within a space into a kind of serial domestic experience. Chapter Five explores how the presence of the car complicates the increasingly fragile sense of home in late-twentieth-century America. Often serving as a literal shelter, it highlights the instability of home, and thus of women's domestic sphere and women's domestic fiction. The final chapter investigates liminality, migration, and "American citizenship." The car, a powerful emblem of American identity, often shapes one's sense of belonging, or not belonging, to American culture, particularly for nonmainstream women. Cars define one's place—or lack thereof—in American society.

This book contends that a fuller appreciation of twentieth-century American culture and twentieth-century American literature results from studying the often overlooked relationship between women and automobility. By acknowledging the extent to which women participate in car culture—and write about it—we can continue to chip away at the increasingly outdated assumptions about women's exclusion from car culture. And given that cars often determine our place in American society, it is particularly important to understand that women's ways of dealing with the age of the automobile are as intricate and varied as men's, if not more so. In America, where the mantra might well be "I drive therefore I am," women have been driving since cars were invented. And in the hands of women writers we get a much clearer sense of how that fact has transformed gender, mobility, and identity itself in an increasingly automotive age.

CHAPTER ONE

Women on Wheels

"A threat at yesterday's order of things"

"Every time a woman learns to drive—and thousands do every year—it is a threat at yesterday's order of things," wrote Ray W. Sherman in a 1927 issue of *Motor* (qtd. in Scharff 117). Given the influence of the automobile on twentieth-century America, it would seem that the presence of the car itself constituted a serious threat to "yesterday's order of things." But Sherman's statement makes a vitally important point: it is not simply the car that has transformed American culture; women's access to it has likewise constituted a revolutionary force for change. When women moved behind the wheel, a new era began. The car, as a powerful machine, may have been initially perceived as the province of men, but women were quick to claim its potential. They greeted the car enthusiastically, even sacrificing various household goods to obtain one. In 1919 a U.S. Department of Agriculture inspector questioned a woman about why her family had purchased a car when they didn't own a bathtub. She immediately responded, "Why, you can't go to town in a bathtub" (qtd. in Reck 8). Automobility is prized over cleanliness—and even over such basics as food and clothing, according to the Lynds' "Middletown" study in the 1920s. "We'd rather do without clothes than give up the car," said one mother. Another woman asserted, "I'll go without food before I'll see us give up the car" (*Middletown* 255, 256). This determination indeed represented a threat to yesterday's order of things—and today's. From its beginnings the automobile has challenged assumptions about the role of domesticity, female responsibility, and even women's identity.

The development of the motor car affected women across the country, helping to break down boundaries between urban and rural life, opening up possibilities to get out of the house and, in so doing, also destabilizing established categories of class and gender. No longer relegated to the home, women now drove into the public sphere, exercising control over the latest technology. Christie McGaffey Frederick remarked in a 1912 *Suburban Life* article: "Learning to handle the car has wrought my emancipation, my freedom, I am no longer a country-bound farmer's wife; I am no longer dependent on tiresome trains, slow-buggies, the 'old mare,' or the almanac. The auto is the link which binds the metropolis to my pastoral existence; which brings me into frequent touch with the entertainment and life of my neighboring small towns—with the joys of bargains, library and soda-water" (qtd. in Berger, "Women Drivers" 57). And once they've tasted soda water, there's no keeping them down on the farm. For women the car provided access to a wider range of possibilities, erasing isolation and changing identity: "I am no longer a country-bound farmer's wife." But this is precisely the problem: once a woman becomes a driver, can she still be a wife, or even a woman?

With women's growing access not only to technological power but also to mobility, new concerns about the stability of the family and the social order surfaced. What happens to domesticity when women are out on the road? Further, what happens to women when they take the wheel? Sidonie Smith argues that "women used automobiles as vehicles of resistance to conventional gender roles and the strictures of a normative femininity" (175), but access to the automobile had a wide range of repercussions for women, both positive and negative. We see a celebration of the new technology and grave concerns about its impact on women's lives and women's power, concerns shared by both men and women, though women certainly have a much less contested vision of what the car has to offer them. The automobile, brimming with contradictory symbolism, is an excellent vehicle through which to explore women's place in early-twentieth-century American culture and literature. Looking at popular consumer culture and literature and their relation to the car illuminates women's complex position in an increasingly technological age. From authors of popular girls' books to Edith Wharton, women wrote their way into automotive culture. Wharton, an extremely astute automotive observer, fully recognized both the possibilities and the liabilities inherent in the power of automobility.

Perhaps more than any other available consumer product, the automobile epitomized the American identity. It offered individuality, mobility, and class status; reflected technological wizardry and good old American know-how; allowed one simultaneously to enjoy and control nature; and provided speed and power to its driver. As automobile historian James J. Flink has observed, "In a culture that has invariably preferred technological to political solutions to its problems, automobility appeared to the expert and to the man in the street as a panacea for many of the social ills of the day" (*Car* 38). Yet the car probably created as many ills as it alleviated, evidenced by the range of contradictory responses to it. The car seemed to herald a new move into the future. But it was also presented as a guardian of the past, advertised as a means of preserving conservative family traditions. Such nostalgia conflicted, of course, with the impact of automotive technology in reshaping human identity, thereby rendering the car a site of the intersection of competing cultural forces and making it impossible to pin down any single meaning for this new and powerful machinery.

In addition to the implicit tension between past and future, the car also suggested the possibility that human identity was, in fact, machine made. On one hand, then, was the excitement of mobility and power; on the other was the growing realization that such power was linked to the forces of mechanical reproduction. The individual autonomy granted by the vehicle began to give way to a sense of mass production, as both people and cars were shaped by the automobile industry, a connection explored more fully in Chapter Two. At the core of these tensions lay the issue of women's access to automobility. In concerning themselves with the relationship between women and the car, contemporary observers both expressed and attempted to control the concerns about the influence of the car—and, in a broader sense, technology itself— on American identity generally as well as on American configurations of gender.

With this range of material and symbolic import, the car posed interesting challenges to literary representation. Several scholars have examined the role of technology in modernist literature;[1] writers were quick to draw upon the metaphorical and stylistic implications of a technological age. "Nothing typifies the American sense of identity more than the love of nature (nature's nation)," says Mark Seltzer, "except perhaps the love of technology (made in America)" (3). The love of automotive technology rises to new literary heights for characters such as Sinclair Lewis's Babbitt. "To George F. Babbitt, as to most prosperous citizens of Zenith, his motor car was poetry and tragedy, love and heroism" (23). The motor car supersedes literature. But the car is more complicated than that—and certainly more than a replacement for art—as Lewis understands, even if Babbitt does not. The automobile serves as a particularly complex literary motif, for it is a material object that combines the power of machinery with a strong sense of anthropomorphism; the car has always been personified, generally as female, possibly growing out of the gendering of ships. The equation of ship and woman seems less problematic; it serves as a home, as protection. The car, on the other hand, gets you out of the house, renders you vulnerable to its own unreliability as well as to the world at large. Thus, the car's female identification comes across as slightly discomforting. It does not offer the same refuge as a ship or a home; rather, the car's gendering (or gender?) resists any easy assumptions about woman, or car, as sanctuary.

The car conveys not just technological power but individual autonomy, granting the driver control over speed, time, and direction. This makes it an apt vehicle for modernist individualism but with a gendered twist: that autonomy was available not only to men but also to women. The automobile, then, provides more than just another opportunity for men to assert control over the female vehicle; it offers women that same mastery and challenges heteronormativity with the consolidation of woman operator and female machine, an issue to which I will return in later chapters. Women writers proved themselves to be well aware of the car's promise and were quick to exploit the spirit of automobility, examining its potential as an instrument of empowerment. The popular media and the burgeoning automobile industry had much to say regarding women's presumed place in the culture of automobility, but, responding to the cultural concerns about gender and power surrounding the automobile, women writers often defied the attempts to use the car as a means to control and domesticate American women.

Women Drivers and the Culture of Automobility

The cultural context surrounding the automobile age reveals that one of the most striking responses to this new form of technology was the car's ability to evoke nostalgia even though, perhaps more than any other object, it signified a move into the future. Yet the car also seemed to open up the possibility of a return to the past, to a simpler life, to the period before the advent of mass transit. A 1919 article in *Harper's Weekly* noted that cars bring the "feeling of independence—the freedom from time-tables, from fixed and inflexible routes, from the proximity of other human beings than one's chosen companions; the ability to go where and when one wills, to linger and stop where the country is beautiful and the way pleasant, or to rush through unattractive surroundings, to select the best places to eat and sleep; and the satisfaction that comes from a knowledge that one need ask favors or accommodation from no one nor trespass on anybody's property or privacy" (qtd. in Gartman 35). This does not sound like a move into a more technologically sophisticated future; rather, it seems like an invitation to "linger," either to prolong the present or return to the past. In some ways the car offered an escape from the future, or at least from the hassles of the present, a concept that still holds sway, according to the Chrysler executive who remarked in 1987: "Escape—that was what a car was. A man's getting away from what and where you are. Get out, and if you don't like your environment or your little life, you get in a car and you can get away from it" (qtd. in Gartman 94). Cars have the ability, in other words, not only to recover the past but also to transform one's identity, to get a person away from "what . . . you are"—at least if "you" are a man. Man is the one who escapes; woman is presumably what he escapes from. This, as decades of feminist criticism reminds us, is a familiar scenario in American literature. Rip Van Winkle's famous twenty-year slumber is initiated by his attempt to escape a nagging wife. At the end of *Huckleberry Finn* Huck decides to light out for the territories, eager to leave behind the ministrations of Aunt Sally and civilization. But what happens when the "you" is a woman and women, traditionally the angels in the house, are expected to be the guardians of the past? Can a woman get away, and how will it transform the shape of American culture if she does? To what extent can one hold onto the past once the woman's sphere has taken to the road?

While rural women may have gained the most from cars, the public debate surrounding women's access to the automobile has focused largely on middle- and upper-class white women. Henry Ford didn't set up his assembly line until 1913, and it took several more years to get the prices down low enough for most working-class whites, African Americans, Native Americans, Latinos, and recent immigrants to buy cars in significant numbers. But as historian Virginia Scharff has argued, the more privileged women still represented a considerable threat: "Their potentially greater access to property, their power as consumers, added a dimension to the problem of public womanhood. Might they use their purchasing power to claim other kinds of authority? Would the woman who felt the jingle of coins in her pocket, and the urge to explore new territory, buy a ticket to an uncharted destination?" (5). Wealthy women have long had purchasing power, but to purchase a car meant more than purchasing clothes, jewelry, or even houses, for cars conveyed a new kind of power: mobility. For a woman to buy a car meant more than cementing her power as a consumer; it meant establishing her authority in a new era of movement and technology.[2] "Women in automobiles entered public space at a time of unprecedented debate over women's right and capacity to step into public life with regard to the ballot box and the university. Such an inroad could not escape notice in a context where the distinction between public and private places served as a boundary defining proper masculine and feminine roles" (Scharff 22–23). The car thus rapidly became a major battleground for women's rights.[3]

Resistance to women's access to the car, of course, gave rise to the myth of the incompetent female driver. A 1904 article in *Outing Magazine,* "Why Women Are, or Are Not, Good Chauffeuses," warned that women's temperament and lack of concentration inhibit their driving skills, though the article grudgingly admits that some women are able to overcome these shortcomings and become decent drivers. Many others were less encouraging; an anonymous writer in the April 5, 1924, *Literary Digest* grumbled, "Some day a man with a head for statistics is going to show us just how many deaths and disablements women drivers are responsible for, and just how much more, or less, dangerous they are at the wheels of motorcars than are their brothers of the road" (qtd. in Boneseal 23). A study by the American Automobile Association cited in *Literary Digest* in 1925 asserted that women drivers have proved "as competent, if not more so, as men" (24). Despite similar studies by insurance companies today, the notion, articulated in 1904, still prevails: that a woman "has to adopt" such skills as motoring, "while a man takes to them naturally" ("Why Women" 155).

Because women have long constituted a potentially major market for the automobile, the auto industry has been reluctant to offend them. To take a hard line that women are dangerous drivers does not help sell more cars. There's been some disagreement over the extent to which women bought cars in the early century: Clay McShane ponders why the number of female car owners was so small, but Scharff argues that the vehemence of the public response to female drivers indicates that a significant number of women were in fact driving.[4] Further, while automakers such as Henry Ford preferred women to remain in the passenger seat, they recognized that women exerted considerable influence over the cars their husbands bought. Roland Marchand cites a 1929 ad in *Printer's Ink* magazine: "The proper study of mankind is *man* . . . but the proper study of markets is *woman*" (qtd. in Marchand 66). Selling cars meant selling cars to women. Thus began a campaign that targeted women as consumer and driver but which walked a very precarious line between urging independence and womanliness. In a 1912 pamphlet entitled "The Woman and the Ford" Ford Motor Company extolled the car as providing an ideal opportunity for women: "It's a woman's day. Her own is coming home to her—her 'ownest own.' She shares the responsibilities—and demands the opportunities and pleasure of the new order. No longer a 'shut in,' she reaches for an ever wider sphere of action—that she may be more the woman. And in this happy change the automobile is playing no small part" (3).

The pamphlet goes on to cite a letter from a woman driver who writes, "There must be women with check books (or access to their husbands'), who love the outdoor life, who crave exercise and excitement, who long for relief from the monotony of social and household duties, who have said, 'I wish I were a man.' Why don't you tell them that your motor car is a solution to all their troubles?" (11). The Ford pamphlet rather deftly dances between offering women what has been largely identified as masculine power and insisting that the car will make a woman "more the woman." The cure, it subtly suggests, for those masculinized and dissatisfied women who want to be men is having their own car.[5] Thus, Ford simultaneously offers to free women from the house and to cement their womanly identity. There is no reason to fear this new technology, suggests the automaker, because women, though they may indulge themselves by playing with it, will remain women; as long as women are holding down the fort of femininity, of home, the "new order" will retain the stability of the old. After all, men still hold the keys—in the form of the checkbook. Even in an article advising women how to operate and maintain a car, Mary Walker Harper advises that the "woman who runs and cares for her own car should dress simply and plainly in a style that will enable her to appear neat and smart under all circumstances," going on to suggest a black satin gown, white hat, and "common-sense" shoes (43). While her advice may be sound, it continues to foreground women's supposed focus on appearance rather than mechanical expertise.

The concern over women's motoring attire may well have stemmed from the uproar that greeted the dress reform at least partly initiated by the advent of the bicycle in the latter part of the nineteenth century. Indeed, the debates over women's cycling anticipated the controversy surrounding female drivers. As Patricia Marks has noted in *Bicycles, Bangs, and Bloomers,* the "rational dress" of the New Woman that "eliminated tight corsets and long, heavy petticoats and skirts, was almost immediately seen as a threat not only to the female image but to male status" (147). The bike may not have been the sole instigator of such change, but it played a major role. In his history of the bicycle David Herlihy points out, "With thousands of women taking to the wheel, bicycling would thus force the issue of rational dress as had no recreational pursuit before it" (267). Scharff opens her history of women drivers with a vision of "the affluent American woman of 1900. . . . Girded in corsets and petticoats and forty pounds of underskirts and overskirts, cloak, and formidable hat, she is clad in immobility" (1). The automobile would erase that immobility. Moreover, the car did not require one to wear a bloomer. Rather, one generally needed to cover one's clothes with an additional leather or rubber duster, along with a veil and goggles. Women auto enthusiasts could thus sidestep the dress debates and thereby challenge the claims that driving made them immodest and unwomanly. Cars, as the Ford Motor Company asserted, made a woman "more the woman"; Ford's language, in fact, echoes the enthusiasm women expressed toward cycling. Maria E. Ward claimed in 1896, "Riding the wheel, our own powers are revealed to us" (qtd. in Garvey 66). Ford, however, offered the car as a cure for boredom and discontent, not as a source of power. But once they were behind the wheel, women discovered that the car's power enhanced their own. The bicycle opened up the gate, and the car roared through it.

Yet while the automobile was changing the world, it was still subtly promising that the gender order, at least, would remain intact. Women may operate the new motor car, but only because men have designed it for them and men have bought it for them. As Laura L. Behling remarks, the auto and advertising industries "created the ideal woman by reinforcing the same traditional sex and gender roles that much of society feared had been irretrievably lost en route to women's enfranchisement" ("Woman" 14). And by preserving ideal womanhood, the industry also defused anxiety regarding the implications of automobility. The new machinery couldn't be too complicated and frightening, if even a woman could drive it. More important, driving did not undermine femininity. In 1909 Alice Huyler Ramsey, accompanied by three other ladies, became the first woman to drive across the country. At the close of her historic journey, the *San Francisco Call* noted, not only did she remain feminine, but she also took a woman's care of the car:

> The drive across the continent by Mrs. Ramsey and her sister motor maids is an object lesson that can not be passed over without considering. Heretofore most of such feats have been performed by women who give the impression of a certain amount of masculine composition in their make up, but in the quartette that arrived yesterday the impression was far different.
>
> From the appearance, outside of a beautiful coat of tan, one would imagine that the car had merely been brought up from Del Monte. It . . . showed that it had received treatment much more considerate than would have been given by man. (Qtd. in McConnell 55–56)

Ramsey's feat, sponsored by the Maxwell-Briscoe Motor Company, constituted a huge publicity campaign partly designed to reduce anxiety over women handling cars; they could do so and still remain "more the woman," keeping the car as tidy as one would expect a housewife to keep her home. Driving had become more a domestic task than an adventure.

In stressing women's ability to drive, most auto companies highlighted the ease of their machinery over women's competence. In another 1912 pamphlet, "Ford: The Universal Car," Ford Motor Company trumpeted the simplicity of the Model T, especially for the woman driver: "On account of the many features for safety in driving and because of its simplicity in control, the *Ford Model T* is especially adapted for the use of the lady driver." "In fact," the pamphlet went on, "there is very little machinery about the car—none that a woman cannot understand in a few minutes and learn to control with a little patience." It's not that women were particularly mechanically inclined; it's that the Ford was so simple that even the feeble-minded should be able to operate it. To reduce female anxiety even further, many women, most notably Dorothy Levitt, published guides for women drivers. Mrs. Sherman A. Hitchcock, in a 1904 column in *Motor* magazine, offered a brief set of directions not only for driving but also for the "mechanical knowledge" necessary to overcome minor obstacles: "If women in general understood the vast pleasure to be derived from driving their own car I feel confident that the number of women motorists soon would be vastly increased. There is a wonderful difference between sitting calmly by while another is driving and actually handling a car herself. There is a feeling of power, of exhilaration, and fascination that nothing else gives in equal measure" (19). This is a very different account of what one could gain than Ford had suggested; rather than becoming "more the woman," a woman could experience "power, . . . exhilaration, and fascina-tion." According to a woman, nothing—neither marriage nor motherhood—could grant a feeling of power equal to that of being a driver.

But such claims were dangerous. And so, having helped to open the door, the automakers carefully tried to rein in some of this might. A 1905 Winton announced itself to be "as simple to run as a sewing machine." The 1917 Ford was "as easy to oper-ate as a kitchen range" (Dregni & Miller 103). The sewing machine, in particular, came up frequently, as Robert Sloss, in 1910, claimed that women viewed cars as "sister organisms" and could handle them better than men: "Shake your head at that all you like, but any man who doubted a woman's grasp of mechanics should watch her fix a sewing machine" (qtd. in Boneseal 24). Such attitudes lingered even into the 1930s, as Priscilla Hovey Wright, in *The Car Belongs to Mother*, contended: "The great differ-ence between the sexes, indeed, is that Man is interested in *how* things work and Woman in how *she* can make them work. . . . Many a vacuum cleaner has been made to live far longer than its manufacturers ever intended it should, simply because Woman with hairpin, razor blade, nail file and embroidery scissors, spurred it into a rally. . . . Man sees the automobile, invented by him, improved by him, fashioned for his pride and pleasure, become, as he believes, the plaything of Woman" (xi–xii). Here Wright masked considerable mechanical expertise by pretending to trivialize women's tools: hairpins and nail files, implements of domesticity. By likening driving cars to women's traditional tasks—sewing, cooking, and cleaning—these ads and ar-ticles reassured women that they *could* drive and men that women would still be in their rightful place: doing domestic labor in the home sphere. Looking more closely at early automobile advertisements uncovers some of the unspoken assumptions and anxieties about women and cars. Ostensibly targeted to women, these ads also clearly reassured men about allowing their women to take the wheel. Advertisements, says historian Gary Cross, "were scripts of social dramas that helped people cope with modern life by giving goods meaning and making them into props that said who con-sumers were or aspired to be" (35). Car ads often propped up increasingly embattled assumptions about gender.

The electric car was aggressively and specifically marketed to women due to its ease of starting and operation as well as its limited range. "No license," Montgomery Rollins claimed in 1909, "should be granted to one under eighteen . . . and never to a woman, unless, possibly, for a car driven by electric power" (279). The identification of the electric as a woman's car reminds us, as Bayla Singer points out, that to speak of "the automobile" tends to collapse important distinctions among cars (32, 37). But the attempt to segregate cars by gender has always proven problematic. Indeed, the demise of the electric car illustrated the perils of a limited appeal. Ford, though it may have targeted women, also insisted that it produced a "Universal Car." Interestingly, the electric was far from being as easy as promised; it was heavy and difficult to charge, and having the battery run down in an isolated spot was far more difficult to remedy than running out of gas. Thus, "the woman who drove an electric might have the illusion of being in control of her transportation, but in fact had less mobility and freedom than she had in a horse-drawn carriage or wagon," explains Singer. "By designating the electric as a 'woman's car,' men at one stroke reinforced their definition of proper femininity and attempted to confine women even more strictly within it" (34).

These dynamics lurk beneath the seeming innocence of an ad for a Baker Electric from *Life* magazine in 1916. The scene evokes the garden of Eden; it positively reeks with idyllic imagery, from flowers to butterflies to a beautifully dressed mother with two little girls, one playing with a doll and one, rather shockingly but innocently, showing her petticoats (Figure 1). Baker's promise to convey the viewer to a social call or the implied suggestion of a day outdoors spells out very specifically that this is not a *serious* machine; it merely conveys upper-class women to their various functions. The heading "Pleasure" reinforces the message: women seek pleasure, not power. The women and girls are clearly not challenging any social conventions, with the only break in etiquette being the attempt to do a headstand. While there is no sign of a chauffeur, indicating that a woman must be at the wheel, driving an electric car—which has a limited range—necessarily keeps one close to home, thereby reducing any threat that this party will head out for the territories.

And yet the symbolism of the scene is striking, with the machine literally in the garden, thus recalling Leo Marx's classic observation that American culture is torn between the machine age and the pastoral ideal.[6] In this advertisement the machine may be posed against an Edenic pastoral, but it is also posed against the figures of women and girls. The question of whether the combined power of women and the pastoral tames the machine or whether the machine takes over the female pastoral remains open, but this scene is clearly meant to convey a soothing calm, a sense that even if the machine has entered the garden, that garden, tended by women, nevertheless endures. Further, the car's open door elides the distinction between the machine and the garden. By opening up the car's interior space to the butterflies, the ad eases the division between interior and exterior and, by implication, between the female and male spheres. The open door provides an escape hatch for the angel in the car even while the car itself limits the extent of that escape. The ad thus offers a safe fantasy for both women and men; women may be able to open the door, but they won't get very far in a Baker Electric: only to the nearest garden.

A 1917 ad for a Liberty Six relies more on text than picture but still focuses on "women's concerns," stressing comfort and safety above power. Titled "When a

Pleasure

—and your Rauch & Lang or Baker Electric is a car of Pleasure.
You find pleasure in the utility by which you so easily reach the out-o'-way places or make a social call—
Pleasure in the ease of control—in the roomy interior, in the genuine coach work, and in the knowledge that your car *is* a Rauch & Lang or Baker Electric.

THE BAKER R & L COMPANY
Cleveland, Ohio

Rauch & Lang Electrics
"The Social Necessity"

Baker Electrics

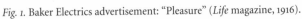

Fig. 1. Baker Electrics advertisement: "Pleasure" (*Life* magazine, 1916).

Woman Drives," the copy assures potential buyers that women will be "infinitely comfortable" and able to stop the car "smoothly" and "safely" with only the "pull of one finger." While ostensibly reassuring women about their ability to handle the car, Liberty reminds them of their natural inferiority: "Where a man might be unconscious of the strength and constant attention necessary to drive most cars, the strain of such driving forces itself constantly upon a woman's mind." Women *could* do it, but it's harder for them—except, of course, in a Liberty. But Liberty wanted to sell cars to men as well, hence the claim, "If a woman can drive a Liberty without effort—all day and every day—think how the Liberty must respond to a man's control." Women could drive the car, but men "control" it. Thus, Liberty subtly reinforces gender stereotypes: women need help, while men exert control. In this ad Liberty also sought to assuage men's technological fears by displacing them onto women; men, too, may prefer the ease, comfort, and safety of a Liberty while claiming the masculine privilege of looking out for the little woman. In an era in which the development of such advances as the enclosed car and self-starter were often derided as dumbing down and/or emasculating the machinery, men could have their technology while not needing to admit to a desire for it.

By 1924 Ford was appealing to the businesswoman with a car to transform a business call into "an enjoyable episode of her busy day" (Figure 2). She may be in business, recognizing financial value and the car's ability "to expedite her affairs, to widen the scope of her activities," but she still seeks a "pleasant car" rather than a powerful one. This solitary woman, however efficient, conveys only a limited sense of power and control at her tidy desk, answering the phone. Significantly, the ad does not picture her actually driving. And another Ford ad from 1924, this one in the *Youth's Companion*, reminds women that they should not get too far above their station in their quest for careers in business; their true role, after all, is being a mother (Figure 3). The ad puts women back in their accustomed place with the children; it is her car, almost a toy, that the boy learns on until he becomes a man and moves on to more powerful machinery. Mother is not teaching Junior to drive; she simply provides the means, not the know-how. Her maternal duty is to buy a Ford. As Ford rather condescendingly informs the boy, "The Ford, as you probably know, is the simplest car made and the easiest of all cars to drive." This, presumably, is why the Ford is Mother's car: it possesses ease and simplicity. Ford joins the child in a conspiracy against the mother; the boy "probably" knows already all about Fords and will go on to surpass his mother in his future "man's" car.

These ads illustrate the promise and anxiety associated with automobility and its implications for transforming gender roles. Clearly, concern over family, technology, and a rapidly changing social order lurks behind the barely concealed hostility to

Her habit of measuring time in terms of dollars gives the woman in business keen insight into the true value of a Ford closed car for her personal use.

This car enables her to conserve minutes, to expedite her affairs, to widen the scope of her activities. Its low first cost, long life and inexpensive operation and upkeep convince her that it is a sound investment value.

And it is such a pleasant car to drive that it transforms the business call which might be an interruption into an enjoyable episode of her busy day.

TUDOR SEDAN, $590   FORDOR SEDAN, $685   COUPE, $525   (All prices f. o. b. Detroit)

Ford
CLOSED CARS

Fig. 2. Ford advertisement: "Her habit of measuring time . . ." (1924). From the Collections of The Henry Ford (87.14.10.18/G2482).

women drivers, even from those trying to sell them cars. If women can maintain this new technology within the old order, then the threat is reduced. Scharff cites *Motor* magazine columnist John C. Long in 1923, doing his best to domesticate the car: "To father the automobile means transportation. . . . Mother sees the car, like the home, as a means for holding the family together, for raising the standard of living, for providing recreation and social advantages for the children" (125). Mother, by Long's account, has no desires of her own; for her the car reinforces family and her own position within it. Long carefully phrases his observation in a way that presents the car as a tool of family life, not a transformation of it. Even father only gets transportation from the automobile, while mother uses it to stabilize and improve the family situation. In other words, the future that the car brings is the same as the past, only better.

But such reassuring messages fell short in the face of simple reality. As Scharff points out, while women "generally did not abandon hearth and duty in favor of the road," they did reconcile "mobility and domesticity, using the automobile to fulfill their family roles as well as to enjoy themselves. In that process of reconciliation they not only transformed woman's place, but also played a subtle yet important part in reshaping the American landscape" (136). Women took to the car and the road in increasing numbers, regendering American culture in ways we are only now beginning to measure. Machinery had always been the province of men, but here was a machine that was both phallic and female. The car was personified as feminine, as evidenced by the famous Model T, the "Tin Lizzie." Early terms for various automotive parts came from women's clothing, such as the *bonnet* (still the standard term in Britain) and the *skirts*, or *modesty panels*, used to hide and protect the inner machinery, thereby suggesting that, even at its mechanical core, the car was a woman. Today we call the protective covers some people put over their front grills "car bras." And to make matters more confusing, the car, as a powerful machine, also remains to a degree masculine. It is masculine in its power yet feminine in being a body that is ridden and mastered.

So what did it mean if women ride and control the car? By driving, women might be perceived as taking over the male prerogative of control and power over a body that is not only feminine but also masculine. Alone with a male machine, who knew what might happen? Michael Berger gets to the heart of the matter: "Another element of the issue concerning women's emotional constitution, though one less openly discussed, was the fear that women would become excited, in the sexual sense, by driving the new machine, or use it as an extension of their sexuality. Many early accounts of women drivers make it clear that ladies often specifically sought out the excitement associated with motoring, a type of excitement formerly reserved for men because only they were thought capable of controlling it" ("Women Drivers" 60). This concern may also have its roots in debates surrounding cycling. As Ellen Garvey notes, those opposed to women on bicycles claimed that the bike's saddle and the vibration of riding "teaches masturbation in women and girls" (qtd. in Garvey 74). While the car's more conventional seat may have reduced the specific "threat" of masturbation, its greater power increased concerns about women's sexual freedom. Even Undine Spragg, the protagonist of Edith Wharton's novel *Custom of the Country*, feels a "rush

Fig. 3. Ford advertisement: "Learn to Drive Mother's Car" (*Youth's Companion*, 1924). From the Collections of The Henry Ford (64.167.19.448/G2480).

of physical joy that drowns scruples and silences memory" in the car (201). Although hardly a self-reliant woman—nor even a particularly sexualized one—Undine nevertheless feels the physical exhilaration of automobility.

One of the most famous advertisements of the era, the 1923 ad for the Jordan Playboy titled "Somewhere West of Laramie," suggests this sexual tension:

> Somewhere west of Laramie there's a broncho-busting, steer-roping girl who knows what I'm talking about.
>
> She can tell what a sassy pony, that's a cross between greased lightning and the place where it hits, can do with eleven hundred pounds of steel and action when he's going high, wide and handsome.
>
> The truth is—the Playboy was built for her.
>
> Built for the lass whose face is brown with the sun when the day is done of revel and romp and race.
>
> She loves the cross of the wild and the tame.
>
> There's a savor of links about that car—of laughter and lilt and light—a hint of old loves—and saddle and quirt. It's a brawny thing—yet a graceful thing for the sweep o' the Avenue.
>
> Step into the Playboy when the hour grows dull with things gone dead and stale.
>
> Then start for the land of real living with the spirit of the lass who rides, lean and rangy, into the red horizon of a Wyoming twilight.   (Dregni & Miller 103)

The ad broke new ground with its evocation of a powerful "broncho-busting, steer-roping girl." Yet the copy is so abstruse as to be almost incomprehensible: "She can tell what a sassy pony, that's a cross between greased lightning and the place where it hits, can do with eleven hundred pounds of steel and action when he's going high, wide and handsome." Presumably, the car represents the pounds of steel and action, a car "built for her." But amid this conglomeration of powerful if confusing imagery, the name of the car stands out: the Playboy. Even without the unfortunate associations currently tied to the term, it nevertheless conveys the aura of play rather than power. And yet, given the "hint of old loves," this car that a woman steps into "when the hour grows dull" also hints at sexuality, "the land of real living." The combination of female driver and the suggestively named male car awakens the possibility of the automobile as sexual partner.

The suggestion does not, however, go too far. The confusing prose discourages close analysis, focusing one's attention instead on a range of seemingly unconnected images.

Moreover, the Jordan Playboy, like the Baker Electric, both liberates and domesticates the woman. Playing with the Playboy displaces the possibility of more overt and forbidden sexual encounters and leaves the driver free to be socialized on "the Avenue," remaining a candidate for marriage and family. The ad's focus is, after all, a car, not a man. While the Baker Electric restricts women to the garden, here we are promised the freedom of the American west—but only symbolically; the driver is offered "the *spirit* of the lass who rides," not her experience. Even the western venue reins in this broncho-busting girl. In 1923 a place "somewhere west of Laramie" is next door to the middle of nowhere, a never-never land. As long as those girls stay somewhere west of Laramie, there's no cause for alarm. In the meantime their weaker sisters can console themselves with the car rather than with sex.

Clearly, these concerns about sexuality are heightened by the bisexuality—and, simply, the sexuality—of the machine itself. The car not only has the potential to transform gender; it also seems to have the power to revolutionize sex, to allow for sexual excitement without the presence of literal male bodies (and, even more disturbingly, also without the presence of figurative male bodies). If sex and gender are radically challenged by automobility, the car enables a radical break from the past rather than a return to it—or even a continuation of it. And in that break lies tremendous potential for women to exercise considerable control over their lives: as wives, mothers, workers, and writers. Women's writing did not, of course, originate with the development of the automobile, but the processes of production and reproduction underwent significant shifts during the Fordist era, with considerable consequences for literature. While Seltzer argues that realist writers such as Frank Norris attempted to replace "female generative power with an alternative practice, at once technological and male" (28), automobile technology did not remain a masculine power. Thus, writers such as Gertrude Stein, Edith Wharton, and Willa Cather were able to draw symbolically from a form of production coded as technological, feminine, and masculine,[7] granting even greater scope for women's literary imagination and production. Even less exalted writers picked up on the car's import, including writers of children's books. These novels may not approach the grandeur of Wharton, but, with their focus on the young, they do offer a fascinating perspective on the training of the next generation of women drivers.

Girls' Books and the Excitement of Automobility

Many writers found expression through the spirit of automobility. Cars burst into fictional texts in short stories, children's literature, and novels, attesting to the degree to which all facets of American culture responded to the automobile age. Ranging from excitement to concern, female voices largely celebrated the advent of the car, especially in the popular press. There is little anxiety displayed in the series of girls' books, written primarily in the 1910s. The *Motor Maids* books, written by Katherine Stokes, the *Motor Girls* series by Margaret Penrose, and the *Automobile Girls* by Laura Dent Crane are cloyingly overwritten, with faulty plot lines and vapid characters. Yet they do offer a fascinating cultural commentary on women's relation to the automobile.[8] The most important, and most obvious, point to make is that all the young ladies who drive cars in these books are precisely that: young ladies who are rich, beautiful, and from the upper class. But not only do they crank engines; they also perform repairs, squirming around under cars with tools and expertise. And while there are certainly young men present, the men rarely take over the wheel or the repairs. At the start of *Motor Girls on a Tour* Cora, the main protagonist, explains why their trip must be girls only: "If we had boys along . . . they would claim the glory of every spill, every skid, every upset, and every 'busted tire.' We want some little glory ourselves" (7). Cora not only declares her ability to deal with such mishaps; she seems to relish the possibility. Part of the appeal of motoring is clearly to demonstrate expertise and control over the car by driving it and maintaining it.

In *The Motor Maids across the Continent* we see Billie, the leader, actually working on her car, crawling under it and "screw[ing] and unscrew[ing] and perform[ing] miracles." That those miracles fail to fix the car seems beside the point; a young woman has mechanical knowledge and is not afraid to apply it. "You may think," says the narrative voice, "that Billie was unusually wise in her generation, but she had had a long training as a chauffeur and could pass muster with the best of them" (83). It's an interesting comment because it encourages us both to believe that Billie was, in fact, unusually wise and also that it is no great shock to find her so. Stokes is thus able to showcase her heroine but also to suggest that it's not such a big deal for a woman to be able to fix a car; it simply takes training. In this way she deftly refutes the claims of women being incapable of dealing with machinery. Alice Ramsey's 1909 transcontinental journey, probably the inspiration for this 1911 story, conveys the same message. The *Cedar Rapids Daily Republican* noted, upon the arrival of Ramsey and her passengers in Iowa, "A number of men, of course, have made the journey, but should Mrs. Ramsey and her three lady companions reach Los Angeles [their destination was in fact San Francisco] as they expect to do, in about thirty days, it will have opened the way for other women, and have proved what Mrs. Ramsey started out to prove, that the combination of a reliable car and a woman who knows it, can make the trip between the two oceans as well as any man" (qtd. in McConnell 41). Women who know cars are thus the equal of "any man." And this knowledge, Stokes and Ramsey assert, clearly lies within women's capability.

But mechanical expertise was not the only source of anxiety in the debates about female drivers. As the advertisements remind us, much of the concern regarding women in cars focused on sexuality: fears that automotive power might convey sexual power, fears that cars offered an easy shelter to copulating couples, fears of breakdowns in lonely places, and fears that young girls could be seduced into getting into strangers' cars. These concerns were not unfounded. Rumor has it that Henry Ford, who was notoriously conservative, tried to design the backseat of the Model T so it would be too small for couples to fornicate in; he failed. In the Lynds' famous study of "Middletown" (Muncie, Indiana), of the thirty girls brought to court for "sex-crimes" in 1923–24, nineteen identified the car as the site of the offense (*Middletown* 258). Ten years later the pattern still held, with a judge telling a women's club, "The automobile is becoming a house of prostitution on wheels" (*Middletown in Transition* 163–64n).

Ramsey, in a possible attempt to defuse such concerns, insisted in an interview published in *Automobile* magazine during her trip: "Good driving has nothing to do with sex. It's all above the collar" (qtd. in McConnell 29). By de-emphasizing the body, she does more than provide the grounds for women as drivers; she subtly claims female identity as intellectual rather than physical—and on an intellectual par with men. This desexing of women drivers and female identity also finds considerable support in the series books, which are remarkably unsexualized. While the Motor Maids' car, called the Comet, is gendered male, the relationship is basically maternal rather than sexual or romantic. When the car is stolen, Billie rebukes herself: "I thought he would be quite safe in this lonely place. It was stupid of me to have left him unprotected like that all night long" (92). She may not have protected him in that instance, but by claiming protection as her responsibility, she does assert a kind of women's power, though hardly a revolutionary one; mothers have always protected their young. Just as the Ford ad reminded women of their maternal automotive duties, so Billie acknowledges her maternal obligation to her car. Yet by positioning the car as the child, she nevertheless claims a more powerful position than that reflected in the ad; she mothers the car rather than the child, claiming responsibility for its existence. In transforming maternity into an automotive force, Penrose situates her young heroine as both producing and sustaining the automobile.

The girls' relationships with real boys, however, are less interesting. Particularly in the *Motor Girls* series, they go no farther than mild flirtation—and not much of that. It's largely good, clean fun, and the few romances are confined to the extra characters, not the primary ones. There is, however, a fascinating scene in *Motor Girls in the Mountains* in which Cora, the heroine, who is lost in the woods, sees a plane land nearby and runs to it for help. She experiences a moment of fear as she approaches the pilot—"Who knew what this man might be? She was alone in this wilderness. Could she trust him?" (137–38). After deciding that "most men were chivalrous," she goes forward to discover that the pilot is a woman—set down in italics and with an exclamation point in the text: "*The aviator was a woman!*" So the fear of violation and danger is raised briefly and then put to rest in the greater excitement over the mechanical expertise of women. Again, what's above the neck seems to trump the vulnerability of what lies below. The pilot, who is not much older than Cora, is trying to beat the record for a flight from New York to Chicago, saying, "There's a lot of satisfaction in beating the men at their own game." "We women all owe you a lot for doing it," Cora replies (139–40). This is a surprisingly feminist tone for a book that, despite its female drivers and flyers, follows a very conservative social line; those to be trusted are primarily from the upper classes and dispense noblesse oblige to the masses, receiving undying gratitude in return. (In fact, a train robber in a *Motor Maid* book vows to go straight following his encounter with the Motor Maids.) All of this adds up to a belief in women's potential—despite a hint of danger and reserved for those who could afford the machines.

These texts are hardly revolutionary. Indeed, their focus on the well-to-do makes clear that there is no intention of upsetting the status quo. The young ladies are protected not just by their automotive expertise but also by their class status. While the car is more than simply a toy, because it does convey some significant power, it primarily functions to showcase the talents of a group of admirable young women, permitting them a wider range for fun, adventure, and good works. The novels encourage their young female readers both to respect class and to view it as linked to ownership of a car. As Behling asserts, the various automobile advertising campaigns reinforced this message, pointing out how a "Body by Fisher" ad implies that to drive a car with a Fisher body is to "move up in class" ("Fisher's Bodies" 516). In Lewis's *Babbitt* a conversation about cars becomes "much more than a study of transportation. It was an aspiration for knightly rank. In the city of Zenith, in the barbarous twentieth century, a family's motor indicated its social rank as precisely as the grades of the peerage determined the rank of an English family" (63). The car replaced the family as a marker of social caste, serving as the means for solidifying class structure rather than challenging it.

Yet gender could not be completely subsumed by class. Comparing these girls' books to the companion *Motor Boys* series by Clarence Young, we can clearly see some limitations placed on the young ladies. Both *The Motor Boys Overland* and *The Motor Maids across the Continent* present a cross-continental journey, but the boys go on their own (though they do pick up various riders en route), whereas the girls take along a chaperone. There are few, if any, girls present in the *Motor Boys* series, whereas the girls always have young men very much in the thick of things. One of the most striking differences is the considerably greater degree of violence in the *Motor Boys* book—from fistfights to gun battles. The boys seem to be engaged in a struggle of good versus bad (presented precisely that baldly) in which good not only wins out but comes with a considerable profit by way of a gold strike. Furthermore, there are simply more car details in this book; we hear more about how the car breaks down, what exactly the boys do to fix it, and their various driving techniques. The girls, on the other hand, while their driving and repair skills are certainly highlighted, seem primarily to be engaged in helping people, in particular helping young lovers. Thus, despite their intrepid journey, their mission is largely domestic—to help the unfortunate to establish homes and families.

These women writers—even of fairly dreadful books—present a much less contested vision of women's power and independence than those actually trying to sell cars to women. It's hard to call them feminist, given the adherence to class hierarchy and the existing social order, but they certainly see no reason why women cannot control the power of automobility. They simply don't see that power as a challenge either to gender—the ladies always remain ladies—or to established tradition. If anything, it reinscribes class status, which makes sense, given the early period in which the series were started, before the car's availability to a wider range of people had significantly increased. The car's revolutionary power is considerably reduced when it functions as a toy of the rich. Yet in their resistance to defining women's identity exclusively through the female body, these writers open a space for women's intellectual achievement—as evidenced through their driving ability. And once women have moved into the driver's seat, the world is all before them.

Edith Wharton and the Liabilities of Automobility

The new world opening to driving women is particularly evident in the work of more complex writers. Edith Wharton, one of the great champions of the motor car, bought cars with enthusiasm, wrote about driving, and addressed the implications of automobility in American life (Figure 4). Unlike Upton Sinclair or Theodore Dreiser, who highlighted the controlling power of the machine and the ways—in Dreiser's 1900 novel, *Sister Carrie*, for example—it seems to eat up the girls who work it, Wharton found this technology liberating and exciting.[9] An ardent advocate of motoring, she purchased a new car with some of the profits from her 1902 novel, *The Valley of Decision*. According to R.W.B. Lewis, when she informed Henry James of her latest acquisition, he wryly replied that with the proceeds of his last novel, *The Wings of the Dove*, "I purchased a small go-cart, or hand-barrow, on which my guests' luggage is

Fig. 4. Edith Wharton in car. Yale Collection of American Literature, Beinecke Rare Book and Manuscript Library. Reprinted by permission of the Estate of Edith Wharton and the Watkins/Loomis Agency.

wheeled from the station to my house. It needs a coat of paint. With the proceeds of my next novel I shall have it painted" (qtd. in Lewis 130–31). James's remarks stemmed from financial envy, not anti-automotive sentiment. He often accompanied Wharton on her motoring tours, thoroughly enjoying the sport. In a 1905 letter to Wharton praising her novel *The House of Mirth,* he wrote, "I wish we could talk of it in a motor-car" (Powers 53). Clearly, by this point the machine had entered not only the garden but also the literary salon. Its move into literary discourse opened up a wider range for teasing out the implications of car culture.

Unlike the girls' books' authors, Wharton conveyed much more fully the symbolic complexity of the car, recognizing its curious tie to both past and future and its problematic relation to female identity and the female body. A writer grounded in the sense of conventional class structure, Wharton saw in the rise of the automobile yet more evidence of its precarious position in the early twentieth century. She subtly depicted the threat to class, order, and identity amid the excitement the car offered. And yet, like so many Americans, she loved it—beginning with her motor tour through France in 1906, which she described in her ensuing travel book, *A Motor-Flight through France,* written in 1908:

> The motor-car has restored the romance of travel.
>
> Freeing us from all the compulsions and contacts of the railway, the bondage to fixed hours and the beaten track, the approach to each town through the area of ugliness and desolation created by the railway itself, it has given us back the wonder, the adventure and the novelty which enlivened the way of our posting grandparents. Above all these recovered pleasures must be ranked the delight of taking a town unawares, stealing on it by back ways and unchronicled paths, and surprising in it some intimate aspect of past time, some silhouette hidden for half a century or more by the ugly mask of railway embankments and the iron bulk of a huge station. Then the villages that we missed and yearned for from the windows of the train—the unseen villages have been given back to us! (1–2)

Wharton's remarks strike a curiously nostalgic note, reminiscent of suggestions cited earlier, a claim that the car would return people to an earlier golden age. Here she seems to share the cultural anxiety over a technological future and thus transforms the car into a means of reclaiming the past, of restoring an old order that had been obliterated by mass transit. But to rediscover the past, the age of innocence, is not to preserve it. Indeed, the automobile would show scant mercy to the unseen villages. On the surface Wharton's rhapsody does not appear to anticipate the extent to which the car would rewrite the future, repave the landscape, and re-gender the country, but her description has some disturbing undertones. The passage is tinged with an almost illicit glee; one "steals" upon a town, "taking it unawares," "surprising in it some inti-mate aspect of past time." The car, not subject to laid-down tracks or schedules, both grants autonomy and privacy to the individual driver and allows that driver to violate the formerly unseen villages, invading their intimate secrets. What is unseen and pri-vate can be private no more.

The automobile takes on the aura of the invader, a term invoked by much of Whar-ton's old New York to refer to the new money and new people infiltrating its society. Clearly, the car carries some baggage that challenges its romance, even to the point of hinting at suggestions of rape, with its violation of intimacy. Wharton, then, shares with many of her contemporaries feelings of ambivalence toward the automobile and its ambiguous connection to gender. By challenging the private realm, the car forces open woman's sphere, exposing it to the public and thus destabilizing the public/private divide so often perceived as gendered, and facilitates women's access to the public world of the motor car. Stephen Kern has observed that the technological advancements around the turn of the century tended to invade the privacy of home; such innovations as the phone and the doorbell, for example, seemed to encroach upon one's personal space (187). "As the public became more intrusive," he writes, "the individual retreated into a more strongly fortified and isolated private world. That is why we can observe in this period both greater interpenetration and a greater separation of the two worlds" (191). Wharton details the influence of the car in realigning the boundaries of these two worlds—and the subsequent cost to women. Celebrating the romance of automobility, her language also conveys an implicit threat to intimacy, privacy, and, in her fiction, female identity.

While some observers, as Berger points out, were worried about women getting too excited by the car, Wharton seems more concerned that women will be either taken over by it or be left behind. At the end of *The House of Mirth* (1905) Lily Bart realizes, "I was just a screw or cog in the great machine I called life, and when I dropped out of it I found I was of no use anywhere else" (319–20). Her inability to adapt to a rapidly changing and increasingly technological world dooms her. Lily's greatest triumph is in the *tableaux vivants* scene, in which she seems to have "stepped . . . into Reynolds' canvas," becoming momentarily frozen in "that eternal harmony of which her beauty was a part" (141, 142). The world of frozen eternity is not the world of the automobile, and Lily dies, unable to draw any sense of possibility from her "rootless and ephemeral" life (331). Another Wharton protagonist, Undine Spragg, however, is made of sterner—and more mobile—stuff. Published only eight years after *The House of Mirth*, the 1913 novel *Custom of the Country* fully engages early-twentieth-century excitement and anxiety about the age of the automobile.

Indeed, yesterday's order of things tumbles without much of a struggle in *Custom of the Country*. This world of our "posting grandparents" is about to be bulldozed to make a highway for the new order; the car that brings one back to the past also obliterates it. The novel details the career of Undine Spragg, a young member of the nouveau riche, trying desperately and ruthlessly to work her way up the social ladder, primarily through multiple marriages to socially prominent men. She comes to New York and catches the eye of Ralph Marvell, scion of the old New York aristocracy, marries him, and destroys him. In this conquest Undine easily eclipses her rival, Harriet Ray, who "hated motor-cars, could not make herself understood on the telephone, and was determined if she married, never to receive a divorced woman" (78). This is the woman chosen by Ralph's mother and sister to ensure continuity and stability. It is significant that Harriet remains unmarried throughout the text. While women may have been expected to preserve the traditions of the past, doing so rendered them unmarriageable, unable to give birth to a new future. If you hate motor cars, the message read, you have no place in twentieth-century America.

Undine, however, represents the new age of automobility—rapacious and unfeeling, highlighting the tension between old and new and, increasingly, between men and women. That Undine does not appear to drive matters little; indeed, Wharton, who had a chauffeur, does not view driving as necessary to claim automotive power. For her the essence of automobile culture lies in the fact of the car's existence and the symbolism of its mobile power. Undine, like the motor car, presides over the demise of the Harriet Rays of the world. Claire Preston has identified Undine as a "buccaneer," a particularly apt label if we transform the ships generally associated with piracy into cars. She functions as an automotive pirate, bearing down on traditional society and appropriating it for her own aggrandizement.

But even before the impact of Undine's explosion into the Marvell clan, Ralph sees his family as "aborigines," primitives threatened with extinction (73). Ralph, caught between the stability of what he terms the "primitive lives" of his mother and grandfather and the allure of Undine's "freshness," is drawn to her "malleability" and her "flexible soul" (82). Undine, in fact, is characterized by mobility; when we first meet her, we are immediately made aware of her movement: she sweeps around "with one of the quick turns that revealed her youthful flexibility. She was always doubling and twisting on herself" (6). This appeal of the mobile reflects the appeal of technology, of speed—of, in short, the automobile. What Ralph fails to take into consideration, however, is the dark side of automotive culture, its revolutionary power to drive through and flatten the old order. If the private world has been characterized by stability and the public world by speed, one might say that Undine speeds into the private and essentially parks her car in the parlor.

In fact, given how completely Undine decimates the old order, one begins to wonder to what extent it ever existed. As Kern comments, "the impact of the automobile and of all the accelerating technology was at least twofold—it speeded up the tempo of current existence and transformed the memory of years past, the stuff of everyone's identity, into something slow" (129). Whereas in *Motor-Flight* Wharton seems to recognize that the car could restore access to the past, in *Custom* she apparently acknowledges that the idyllic past may itself be a product of automobility. Ralph's identification of his family as aborigines marks them as anthropological objects rather than living characters. Identified in the novel with such static objects as houses, portraits, jewelry, and tapestries, the past never seems to have been alive and only gains value in retrospect as heirlooms, artifacts. Even that value appears suspect as Undine resets Ralph's heirloom jewelry and tries to sell Raymond's tapestries. Not only is the past dead and devalued; it is also for sale. In her consideration of the novel Debra Ann MacComb has noted that "a contemporary trend in American advertising that emphasized benefits derived from the consumption of a product rather than the product itself" equated the "process of consumption" with "social survival" (768). Artifacts such as tapestries, which are not to be "consumed," thus lose worth except in terms of the money for which they can be exchanged. In contrast, the automobile functions as both a consumable good and a consuming one, given the gas, oil, and maintenance necessary to operate it. Hence, it far surpasses items of traditional worth such as art and jewelry, privileging technology over heirlooms. In an automotive society in which the movement and exchange of goods command value, the present and future belong to those on the move.

The ones on the move were, surprisingly, often women. We may not see too many literal women drivers in Wharton's fiction, but Undine evokes such women as Alice Ramsey, whose feat resonated among the female populace. Ramsey reports having passed a woman waiting on the side of the road in Iowa, who stopped her car and said: "I read about you in the paper and I've come six miles to see you and I've been waiting for a long time. Yes, I'm sure glad I saw you!" (42). Clearly, even the sight of a woman at the wheel empowers other women. Undine, of course, though desiring the power and autonomy displayed by Ramsey, lacks her self-sufficiency, being tied to the traditional assumption that it is man's obligation to provide women with comfort and wheels. Nevertheless, she challenges any notion of a stable and fixed female sphere in her need to be constantly on the move, evident in her bodily restlessness, relentless activity, and multiple marriages. Although she resents not having a car of her own and being dependent on friends for lifts, she nonetheless embodies the spirit of the automobile in her refusal to acknowledge any restraint or imposed obligation. As she heads off for a joyride with the married Peter Van Degan, conveniently forgetting her son's birthday party, the car offers her a release from responsibility: "as the motor flew on through the icy twilight, her present cares flew with it" (202). Just as Wharton reveled in being able to escape the bondage of the train, so Undine breaks free of the social dictum that says that woman's place is in the home—her husband's home.

The closest she comes to such imprisonment is during her marriage to the French aristocrat Raymond de Chelles, when she languishes on his country estate, where all movement is by horse and carriage or train: "The days crawled on with a benumbing sameness . . . and occasionally the ponderous pair were harnessed to a landau as lumbering as the brougham, and the ladies of Saint Désert measured the dusty kilometres between themselves and their neighbors" (511). Wharton's careful choice of words here—*ponderous, lumbering*—highlights the slow pace of life that comes closer than anything else to breaking Undine, threatening even her female identity, as evidenced by her inability to bear Raymond a child, a calamity that her mother-in-law attributes to the extended wedding journey: "Who could tell, indeed, whether these imprudences were not the cause of the disappointment which it had pleased heaven to inflict on the young couple?" (512). Too much movement, the comment plainly suggests, can make a woman infertile.

The Marquise was not alone in such concerns; women, weak and fragile creatures, were not supposed to go dashing off from place to place. Such behavior led to questions about the effects of the car, the primary vehicle for such mobility, on the female body. Indeed, there was considerable medical concern, particularly for women, regarding the safety of this new means of travel. In a paper delivered to the New York Obstetrical Society in 1911 Dr. J. Clifton Edgar carefully detailed his observations on the effects of motoring on various bodily ills: "Inquiry amoung [sic] my patients points to the fact that at first the use of the motor car in a constipated subject secured a daily evacuation of the bowel. . . . But as the automobile was used more continuously the effect, like that of many laxatives, wore off, and the original constipation returned. . . . Broadly speaking, the continuous use of the motor car is, in the long run, bad for the bowels" (1001). He spent a great deal of space detailing the ways in which motoring affects the nerves, concluding that those who tended to be nervous may be made worse by driving, but women who were not prostrated by nerves should be able to drive without incident. The car, not requiring significant bodily exertion, thus provided an ideal means for women to gain air and exercise; indeed, Wharton's Undine, in the throes of her "nervous breakdown," requires Ralph to hire a car to take her on daily drives.

But the biggest medical concern about women in cars was in the area of reproduction. Responding to a widespread belief that cars caused miscarriages and sterility, Dr. Edgar argued "that the unfavorable influence of the automobile upon pregnancy has been somewhat exaggerated" (1003). While he recommended care, moderate speed, and short journeys with frequent breaks, he felt that most women could drive throughout pregnancy without experiencing adverse effects. They could drive throughout pregnancy but needed time in order to get pregnant, something that Undine does not seem to have, being too busy moving to spend time in backseats. Indeed, her problem is not that she spends too much time in the car; it is that she *is* the car. Rather than employing the car, she seems to embody it, for the automobile waits for no one; it demands accommodation, change, attention, and money. As Paul Frankl, an American auto designer, remarked in 1932: "Twentieth-century Man is in the toils of a new mistress . . . the Machine. . . . Roughshod she trampled over all traditional values; she crushed out the lives of men, women, and children; she destroyed the old beauty and replaced it with a 'new ugliness'" (qtd. in Gartman 56). Ugliness aside, this new mistress sounds like a dead ringer for Undine Spragg Moffatt Marvell de Chelles Moffatt, destroying the old order and driving through the lives of men, women, and children. More sexually active than Lily Bart but, interestingly, less sexy, Undine is almost mechanical in her rapacious quest for upward mobility, evincing little more feeling for her victims than the automobile does for roadkill. Undine is not so much the driver as the car itself, reflecting Wharton's concern not about women getting sexually excited by the automobile but of automobility defining women's bodies and women's identities.

Her concern was well founded. The association of the female body with the automobile has been around almost since the first car was designed, despite Alice Ramsey's attempt to separate women's driving ability from their bodies. Interestingly, the intersection of men's bodies with the car generates less anxiety. George Babbitt feels empowered by his association with the car: "He noted how quickly his car picked up. He felt superior and powerful, like a shuttle of polished steel darting in a vast machine" (Lewis 45). The sexual imagery implicit in the passage suggests that the car enhances Babbitt's potency. Women's bodies, however, are more likely to be presented as sexually objectified than sexually active. The designer Norman Bel Geddes created for General Motors a "Futurama" for the 1939 New York World's Fair, focusing on the need for superhighways. It was the fair's most popular exhibit, seen by nearly twenty-seven million people. He also designed, for the same fair, a peep show called "Sexorama" featuring seminude women dancers surrounded by mirrors, "a machine-age dance that made a woman's body seem infinitely replicable like the interchangeable parts on the assembly line" (Rydell 139). Babbitt may feel "like a shuttle of polished steel," but these women are replicants, assembly line products. Futurama and Sexorama—cars and women. Does the association spell change or more of the same? ·

Wharton's opinion is divided, for while Undine apparently views husbands as infinitely replaceable and interchangeable, her willingness to have her life defined by the spirit of technology may also anticipate the day when she herself becomes an assembly line model. If cars control the culture, money buys the cars, and money, in Wharton's fiction, is usually controlled by men. Like the female letter writer in the Ford pamphlet, Wharton realizes that women need access to their husbands' checkbooks in order to exercise automotive power. Undine may select and discard husbands with the aplomb of the rich trading in cars, but like the Model T and unseen villages of our posting grandparents, she is rapidly becoming obsolete. Simply gaining mobility does not in itself automatically transform women's lives—as any carpooling mother can attest. Wharton recognized that one needs more "above the collar" than beauty.

Ultimately, Undine's reliance on her physical appearance marks her as still trapped by the female body, a trap that Wharton details assiduously. Unlike the girls' books' authors, she recognizes that it may not be so easy to assume that beautiful women's bodies are only enhanced by automobility. Martha H. Patterson has suggested that Undine represents Wharton's "feminization of larger cultural anxieties over unregulated corporate power," functioning as "proprietary capitalist, commodity, and dynamic of exchange" (213, 226). Unregulated and commodified, Undine is clearly slated for a life of continual disappointment because her desires are ultimately insatiable: "Even now, however, she was not always happy. She had everything she wanted, but she still felt, at times, that there were other things she might want if she knew about them" (591). But what she represents—the power not only of automobility but also of the corporate world that controlled the automobile industry—is far more difficult to dismiss. Having helped to set these forces in motion, she is about to be run over by them.

This novel, with its surprisingly unlikable protagonist, reflects the power of consumer culture and the conspicuous consumption made famous by the economist and social critic Thorstein Veblen. Rita Felski suggests that in the modernist era the "emergence of a culture of consumption helped to shape new forms of subjectivity for women, whose intimate needs, desires, and perceptions of self were mediated by public representations of commodities and the gratifications that they promised" (62). Certainly, Undine, identified by Preston as "the first truly modernist female protagonist" (141), defines herself through commodities. Her desires are less sexual than material. As Felski notes, the "discourse of consumerism is to a large extent the discourse of female desire" (64–65). Yet what is being sold in this novel goes beyond consumer goods; what is being sold is the construction of women as mobile beings, as drivers, and, ultimately, as cars. Just as the automobile reconstructs and erases the past, so does Undine reinvent herself after each divorce, conveniently forgetting earlier husbands in her attempt to present herself as a new object in the marriage market. MacComb discusses the novel in the context of the growing divorce industry, which creates a "product—marketed in terms of the increased freedom, mobility, and status it can provide—that keeps the marriage economy expanding because spouses and even families become disposable items in the rotary system of consumption" (771). Divorce and the automobile both construct mobile women who are ultimately "disposable" in an assembly line economy.

Custom of the Country, published in the same year that the first cars rolled off Henry Ford's assembly line, reflects a blind dash into an unknown future. With the advent of mass production almost anyone could buy a car, and almost everyone did. Wharton anticipates what the girls' books' authors did not: cars challenge the existing social structure. Undine's final marriage, which seems to exist only in motion—"They were always coming and going; during the two years since their marriage they had been perpetually dashing over to New York and back, or rushing down to Rome or up to the Engadine" (576)—may soon be the norm rather than the luxurious exception she seeks. Riding roughshod over the obstacles in her way, she gains "more than she had ever dreamed of having" (591), exercising the control and power of a woman at the wheel. Yet amid the mobility and excitement remains a troubling question: that of human identity now that the machine has entered not only the garden but the house—and the body—as well. Undine's inability to stop and appreciate what she has reflects the combustion engine's constant need for refueling; like Undine, it has no meaning at rest.

Concerns about human identity in a changed era help to shape Wharton's fiction as she documents the loss of custom, innocence, and possibly femininity to the age of the automobile. Her presentation of Undine's almost mechanical character reflects American cultural anxieties regarding the influence of the car, even as America (and Wharton) celebrated its excitement. The allure of the past was giving way to the energy of the present. Particularly striking is the extent to which these concerns were articulated through women's bodies, whether in advertising, popular literature, or "high" culture. Baker Electric may have employed beautifully dressed ladies to defuse anxiety over the motor car's revolutionary new power, but, as Wharton reminds us, one could no longer rely on the ladies to maintain order and stability. And despite Alice Ramsey's claim that motoring skills depend on what's above the collar, women's bodies were firmly ensconced as part of the symbolism of the automobile industry. The Tin Lizzie, often called "Henry's lady," remained the icon of the automotive world for decades.

Such cultural upheaval could not help but be reflected in the literature. As the *Motor Girls, Motor Maids,* and *Automobile Girls* books encouraged young women to hit the road in search of adventure, writers such as Edith Wharton both seized upon this potential and cautioned readers about where it might lead, particularly in its ability to reinforce and unmake class. In her fiction Wharton initiated new patterns for women's literature: the focus on the individual over the community, autonomy over marriage, and the gradual disintegration of the family. Despite cultural attempts to use the car to keep women in their place, Wharton's work reveals that if separate spheres ever did exist, the car had begun to bridge that gap, as Wharton herself proved to be among the first American women writers to bridge the narrowing distance between men's and women's literature. She stands as evidence that women have moved into the literal and literary driver's seat, that they have begun writing behind the wheel.

Amid all the excitement and romance attendant upon the advent of the car, Americans clearly felt a lurking anxiety regarding its possible consequences. Edith Wharton may have been one of the more astute observers, but she was by no means the only one to note its contradictory implications. Redefining the past and refiguring the future, the car invoked nostalgia for an era that may never have been while at the same time both promising and challenging the continuation of that fictitious time when gender and family were stable entities. Its coming helped to chip away at the increasingly damaged divisions between the male and female spheres, bringing women a symbolic, if not actual, vehicle for social, political, and economic advancement. Slow and cautious as that advancement was, reined in by the marketing and cultural representation of the motor car, it nonetheless provided a foundation to question the role of women and the meaning of *woman* now that American women were on wheels. Despite the attempts to curb women's power in car culture, women's access to automobility was in gear and on the move, threatening to leave behind home, family, and the traditional duties of the private sphere.

CHAPTER TWO

Modernism

Racing and Gendering Automobility

In no era of American life was the automobile more significant than during the period of modernism. Automobile production rose from 4,000 in 1900 to 187,000 in 1910 (Rae 33). By 1914, the year after Ford began assembly line production, sales of the Model T alone reached nearly 270,000 (see Rae 61). By the 1920s the auto industry was the largest in the nation; by 1927 the United States was home to 80 percent of the world's motor vehicles, one for every 5.3 people in the country (Flink, *Car* 70). In their famous study of "Middletown" in the 1920s, the Lynds discovered just how important the car had become to people's lives, as they found some residents mortgaging their homes to buy cars (*Middletown* 254–55). "'Why on earth do you need to study what's changing this country?' said a lifelong resident and shrewd observer of the Middle West. 'I can tell you what's happening in just four letters: A-U-T-O!'" (qtd. in Lynd 251). When then-president of Princeton, Woodrow Wilson, warned in 1906, "Nothing has spread Socialistic feeling in this country more than the use of the automobile" (qtd. in Gartman 15), he was concerned over class envy for those who owned these luxury vehicles. But the advent of the assembly line opened the door to mass consumption of what was rapidly becoming a necessity rather than a luxury. Indeed, in his 1928 nomination acceptance speech Herbert Hoover proclaimed his vision of *two* cars in every garage, a remarkable development in just over twenty years (Rae 106).

The sheer immensity of the numbers, combined with the enthusiasm of the response, affirms the hold the automobile had on modern America. It provided economic opportunity—and often incurred economic dependence—as well as inspiring love and loyalty. With this degree of influence the car could not help but shape modernist ideology, contributing to the transformation of the natural body into the technologized body and thereby destabilizing the link between gender, race, and the physical body. Considering the role of the car in modernist literature opens up new ways of conceptualizing race and gender in modernism and challenges some of the divisions between technology and the body, between high art and mass culture, between masculinity and femininity, and between white and black that so often seem to define our understanding of the period.

Because the automobile plays such a crucial role in a time of great cultural upheaval, we must consider it within a very broad spectrum, beginning with what many would identify as the core of modernist identity, the white male body. Marlon Ross has observed: "To be a modern, straight-up, quick-thinking, forward-looking U.S. citizen ready to compete in the new century is to carry the self—to boast, stare down, and swagger, for instance—in ways commonly seen as off-limits not only to women but also to nonwhite men. In other words, the signs of self-confident modernity are marked on the body in ways normally defined not only as white but also as masculine" (50). Yet if such signs are marked on the body, then how does the car's role in shaping that body affect the construction of white masculine modern identity and, by extension, nonwhite masculinity and femininity? How can one "swagger" in an automobile? The answer, which I explore in this chapter, is "very easily"—but with a great deal of attendant anxiety. Women and men writers, white and black writers, all project onto the car a host of modernist concerns underlying the position of technology in constructing the body and the self. At the core of those anxieties one finds questions regarding race and gender, though not all writers explore both categories. The origins of such questions can be traced back to a largely forgotten automobile race, in which anxieties regarding race, cars, and masculinity were played out. The double meaning of *race* speaks volumes in regard to the importance of the automobile in unsettling racial politics; it both reifies and destabilizes white privilege, enforces and reshapes masculinity. This race seized the public imagination because it pitted a white man against a black one who had already proved that masculine power was not restricted to white men.

In 1908 the color barrier of the winner's circle in professional sports was breached for the first time when African-American boxer Jack Johnson defeated Tommy Burns to become the world's heavyweight boxing champion. A huge brouhaha ensued. Boxing, the sport of the ancient Greeks and Romans, the field in which masculine physical prowess is established, suddenly seemed to betray its white proponents regarding its ability to define the pinnacle of masculine bodily achievement. Because, as Michael Berger points out, "the championship was seen as the embodiment of masculinity, a member of an 'inferior' race should never be given the opportunity to hold it" ("Great" 59–60).[1] What made Johnson's victory particularly galling to white America was his flamboyant, in-your-face behavior, as he publicly flaunted his gold-capped teeth and penchant for copious and expensive liquor, fast cars, and white women. Unlike Jackie Robinson forty years later, Jack Johnson did not refrain from answering back. In an attempt to challenge Johnson's title, an elaborate claim, eagerly supported by the writer Jack London, was devised that because former champ Jim Jeffries had retired undefeated in 1905, voluntarily relinquishing the title to Marvin Hart, who was then beaten by Tommy Burns, Jeffries, not Burns, had technically still been the champion. Because Jeffries had never been defeated, the argument went, he still owned the title and was simply on temporary sabbatical. Jeffries was pressured into returning, five years after his retirement, to face Johnson. As Jack London pleaded in the *New York Herald:* "Jeffries must emerge from his alfalfa farm and remove that smile from Johnson's face. Jeff, it's up to you!" (2.3).

Jeffries returned reluctantly, hoping, he said, to justify the hopes of "the white race that has been looking to me to defend its athletic superiority" (qtd. in Blassingame 3–4). But Johnson knocked him out—the first man ever to accomplish this feat. Although others claimed that Jeffries, aging and out of shape, would have defeated Johnson five years earlier, after the fight he himself admitted (after first agreeing that he was too old) that such claims were unjustified: "I could never have whipped Jack Johnson at my best" (qtd. in Farr 119). This second Johnson victory dealt a serious blow to boxing as a sport. Suddenly, it seemed less popular. As the *Chicago Tribune* noted, "It is apparent *now* that prizefighting is an ignoble pursuit" (qtd. in Berger 62; my emph.). It had only become an ignoble pursuit, in a society grounded in notions of white supremacy, once an African American could triumph in it. But Jack Johnson did not rest on his laurels. An amateur car racer, he took the contest into a new realm and challenged Barney Oldfield, the auto racing champion, to a car race.

And this is where, for my purposes, the story gets really interesting. The race between Johnson and Oldham, which took place in 1910, sets up a number of intersecting discourses surrounding race, automobiles, and masculinity. If a member of an "inferior race" should never be given the opportunity to hold a boxing title, what about a car racing title? To what extent does driving a car replace the now ignoble pursuit of boxing, and what implications does African-American participation hold for auto racing—and car culture? One must remember that while we may no longer think of auto racing as an integral part of current automobile culture, it was crucially important in supporting the early automotive industry. Henry Ford first made his name as a race car driver, though he quickly passed that job on to none other than Barney Oldfield. It was this success at the track that encouraged the original investors to take a risk on the Ford Motor Company in 1903. If Henry Ford could build and drive a successful race car, then maybe he could establish a successful automobile company. Participation in auto racing thus meant potential impact in shaping the growing automobile age. And shaping the age of automobility shaped twentieth-century America. As Charles Wilson, chairman of General Motors, would later so famously tell Congress, "What's good for GM is good for the country." In some ways the car *is* America.

All this adds up to a considerable stake in keeping Jack Johnson out of the auto racing business. And yet, given his physical prowess, the only way to put him back in his "place" seemed to be by car. As the *Philadelphia Inquirer* noted: "The only thing that can beat Jack Johnson is an automobile. Judging from past performance, the machine stands every chance of taking up the white man's burden" (qtd. in Berger 59). Barney Oldfield, billed as the "Great White Hope," did, in fact, handily defeat Johnson in two heats, though the circumstances surrounding the race were anything but normal. The American Automobile Association (AAA), the governing body of the new sport of auto racing, refused to sanction the race, claiming that Johnson's "entrance into the sport would be detrimental to its best interests" (qtd. in Berger 63–64). Some of the alleged reasons included Johnson's lack of racing experience—despite the fact that he had participated in several private matches and done reasonably well. It's difficult not to speculate that racial identity had something to do with it. After all, if an African American's success at boxing could render it an "ignoble pursuit," the AAA may well have feared that auto racing would face the same fate, and the organization later suspended Oldfield for his participation in this outlaw event. Oldfield was both egged on to defend white supremacy and rebuked for failing to keep car racing pure— in many senses of the word—and for turning the event into a media circus that highlighted race rather than cars. The automobile weekly *Horseless Age* announced that "B. Oldfield has embarked in the coal business in colossal style . . . he is catering especially to that class of 'sporting persons' interested in such elevating pastimes as cakewalks, craps, ragtime, chicken dinners and policy" (qtd. in Berger 65). Although many spectators were disappointed that Oldfield had not lured Johnson into an accident with the hope that this could render him physically incapable of defending his boxing title, there was great relief at the outcome, the headline in the *New York Sun* reading, "OLDFIELD SAVES WHITE RACE."

To "save" the white race, one needed the machine that could take up the white man's burden: the automobile. As the example makes clear, the car not only enhanced personal mobility; it challenged the hierarchies of race and gender and the social positions such hierarchies imposed. In so doing, the car helped to mold the modern age. Regardless of Oldfield's victory, the implications of Johnson's automotive challenge lingered. One could not count indefinitely on Oldfield as savior, particularly given that he was suspended from racing for agreeing to take on Johnson in the first place. Johnson's loss at the racetrack may, ironically, have proven more devastating to white supremacy than his success in the boxing ring because it heightened awareness that cars were changing the very groundwork of white male identity. Paul Gilroy observes:

"Johnson's place in the bitter political conflicts over manliness, masculinity and 'race' that greeted the century of the colour line have been considered at length by other writers but their accounts have not usually been engaged by his inauguration of these distinctive patterns in the way that automobiles were understood and articulated within and sometimes against American racial codes. Johnson's manliness and his ambivalent stardom were both expressed in and through his command of the cars which drew hostility, harassment and introjected, covetous admiration from the police wherever he went" (99). He may have lost to Oldfield, but he continued defiantly to drive fast cars, challenging the racial codes that associated automobility with whiteness. For Johnson the car may have meant more than even boxing, as his biographer Randy Roberts notes. At his trial for violation of the Mann Act in 1913 he asserted, "My mind is constantly on automobiles" (qtd. in Roberts 75). By insisting that black men could drive, he reshaped what it meant to be a man.

The car's overt role in making modern men has obvious implications for making modern women. Exploring the gender wars surrounding the advent of the motor car, Clay McShane explores the extent to which men guarded cars and automotive technology from female encroachment. In challenging male automotive dominance, women challenged masculinity itself: "It is important to remember that cultural definitions of gender are synergistic; i.e., when women seek to redefine their roles, implicitly they redefine men's roles and vice versa" (McShane 151). Given that masculinity was increasingly defined against femininity, one must consider the automobile's influence on men in order to understand the full extent of what it meant for women to align the body with the machine. Manhood had long been defined as the opposite of childhood, says Michael Kimmel: "At the turn of the century, *manhood* was replaced gradually by the term *masculinity* . . . contrasted now with a new opposite, *femininity*" (119–20). Mark Seltzer in *Bodies and Machines* suggests that at the same period the closing of the frontier resulted in "a relocation of the topography of masculinity to the surrogate frontier of the natural body" (150).

But if masculinity was now tied to the natural body, that body was increasingly challenged by technology. It was, after all, not Oldfield's body that defeated Johnson; it was Oldfield's car and his skill as a driver. This complicates the rather easy assumption that masculine prowess is determined by who knocks down whom; now one needs access to the right car and a different kind of talent.[2] Masculinity—and human identity—is no longer solely grounded in the human body; in the modern era it is inextricably linked to technology and to the automobile. "In the throes of massive economic and social change," writes McShane, "men defined the cultural implications of the new automotive technology in a way that served the needs of their gender identity. In their efforts to exclude women from control of the new technology, automobiles were again more than just machines" (149). Automotive technology shaped gender.

Technology is integral to modernism, whether one argues that the modernists resisted machine and consumer culture or embraced it. Cecelia Tichi, examining the impact of technology on modernism in *Shifting Gears,* notes that "the gear-and-girder world existed in a dialectic relation to another, much darker world in which nature and civilization were subject to flux and upheaval. This was the world of instability" (41). The "gear-and-girder world" of scientific stability intersects with what Tichi rightly identifies as a "much darker world," a raced world, I would argue.[3] But as Jack Johnson made clear, that darker world also drives, thereby challenging the association of progress and science with whiteness. Nature and civilization are in flux when non-white people begin to challenge white hegemony, a challenge facilitated by the automobile. Thus, while the machine may help one to take up the white man's burden, it also destabilizes the hierarchical social structure that supports white supremacy by adding new avenues for mastery and success.

Further, just as the machine age seems to de-emphasize the power of the body—just as Barney Oldfield in a car can both defeat Jack Johnson in a car and symbolically devalue Johnson's bodily prowess as a boxer by trumping his win in the boxing ring with a loss at the racetrack—the growing emphasis on the machine age highlights the "natural" body as in danger of displacement yet refusing to quit the stage. We may defeat Johnson's body, but that defeat certainly doesn't erase Johnson's body. Seltzer argues that "those more visibly embodied figures [by which he means the female body and the racialized body] are the figures through which these tensions can be at once recognized and displaced or disavowed" (64). Or, as Tim Armstrong declares in *Modernism, Technology, and the Body,* "Modernism, then, is characterized by the desire to *intervene* in the body; to render it part of modernity by techniques which may be biological, mechanical, or behavioural" (6). Modernity is also marked by the growing intertwining of the machine and the body. In particular, modernity can be characterized by the ability of the most popular machine of the day, the automobile, to annex the human body.

And both technology and modernity are, of course, strongly linked with masculinity. Rita Felski, in *The Gender of Modernity,* has noted the feminist argument that "such phenomena as industry, consumerism, the modern city, the mass media, and technology are in some sense fundamentally masculine" (17). Yet the symbol of the car unsettles this gendering, for the automobile is both phallic symbol and female object. Thus, when the car became the site of modern masculinity, it revealed the fragility of male identity and privilege in an increasingly technologized culture. The automobile played a crucial role in shaping the modernist tensions between the machine and the body; it both displaced and displayed the body, replacing some bodily functions yet demanding bodily skill. It represented both masculine and feminine power. And more than any other machine, it became anthropomorphized in American culture, generally functioning as both extension of the self and treasured companion, both subject to human control yet often seeming to possess a mind of its own.

The automobile industry itself is a prime example of these tensions regarding the place of the physical body in a machine age. Henry Ford is well-known for his role in establishing assembly line labor, producing cars with interchangeable parts and workers with interchangeable skills and even identities. Ford's vaunted five-dollar-a-day wages, established in 1914 as a significant jump from the previous average of roughly two dollars a day, provided him with the opportunity to enforce his vision of what a man should be. The salary came with strings attached. For one thing, it was restricted to men. They had to have been working for the company at least six months, be over the age of twenty-two unless they were married or supporting a widowed mother, and as Ford put it, "The man and his home had to come up to certain standards of cleanliness and citizenship" (*My Life* 128). In order to judge cleanliness and citizenship, Ford sent inspectors from his Sociology Department out into Detroit to gather information on his employees. They asked, among other things, about marital status, religion, citizenship, savings (including passbook number), value of house, hobbies, number and ages of children, health, and name of the family doctor.[4]

In monitoring the conditions of his men, Ford, the great automotive engineer, was experimenting with human engineering. All foreign-born employees were required to take English-language courses after working hours, taught by American-born workers for no pay. The graduation ceremony featured a large cauldron bearing the sign FORD ENGLISH SCHOOL MELTING POT—a literal melting pot, which employees entered wearing the garb of their native countries and emerged from dressed in American clothes, carrying American flags (Lacey 126). Both scenes—of the inspectors' surveillance and the graduation ceremonies—are brilliantly portrayed in Jeffrey Eugenides's 2002 novel *Middlesex,* in which the graduates "rise from the cauldron. Dressed in blue and gray suits, they climb out, waving American flags, to thunderous applause," while the orchestra plays "Yankee Doodle" (105). The whole procedure is eerily reminiscent of Hank Morgan's man factory in Mark Twain's novel *A Connecticut Yankee in King Arthur's Court.* In fact, Ford claimed that, given "the most shiftless and worthless fellow in the crowd," by means of giving him a job with a decent wage and hope for the future, he could "guarantee that I'll make a man out of him"(qtd. in Lacey 127). As Antonio Gramsci later wrote in his *Prison Notebooks,* Americanism and Fordism constituted "the biggest collective effort [ever made] to create, with unprecedented speed, and with a consciousness of purpose unique in history, a new type of worker and of man" (215). In other words, Ford created the mass-produced man, efficient and inter-changeable.

The process of making men went hand in hand with making cars. "A business," Ford said, "is men and machines united in the production of a commodity, and both the man and the machines need repairs and replacements" (*My Life* 159). "There is every reason to believe," he added elsewhere, "that we should be able to renew our human bodies in the same manner as we renew a defect in a boiler" (*Philosophy* 12). He made it clear that not only were men interchangeable with one another; they were also interchangeable with machines and required only mechanical repair. This vision is not unique to Henry Ford. As Seltzer notes, if "turn-of-the-century American culture is alternatively described as naturalist, as machine culture, and as the culture of consumption, what binds together these apparently alternative descriptions is the notion *that bodies and persons are things that can be made*" (152). Thus, the construct-edness of human identity was not so much in question as was the nature of that con-struction: *how* one was made. Some of Ford's closest associates picked up on this, re-gretting that Ford himself was not put together as well as he might have been. Samuel Marquis, former director of Ford's Sociology Department, lamented in 1923 that while Ford was "a genius in the use of methods for the assembly of the parts of a machine, he has failed to appreciate the supreme importance of the proper assembly, adjustment, and balance of the mental and moral machine within him. He has in him the makings of a great man, the parts lying about in more or less disorder. If only Henry Ford were properly assembled! If only he would do in himself that which he has done in the factory!" (qtd. in Banta 273–74). While challenging Ford's own assem-blage, Marquis fully acceded to the Fordist philosophy on the making of men—it was an assembly line procedure.

Meanwhile, Alfred P. Sloan, along with his style maven Harley Earl, was marketing image and planned obsolescence over at General Motors with such success that by 1927 GM topped Ford as the number one automaker. As David Gartman points out, "Earl and his ilk understood well that the automobile was valued more for its sym-bolic than its functional power, as testimony to progress and success in the consump-tion world insulated from Fordist production" (117). Shifting the focus from engi-neering and cost to style suggests a move from body to image, from material to symbolic. Yet the emphasis on styling, on finding a car that matched one's identity, also subtly reinforced the emphasis on the body. General Motors's highlighting of beauty and design reminds us of what bodies are supposed to look like. By masking standardization—GM cars, of course, also came off assembly lines—Harley Earl granted one the illusion of individuality and escape from mass culture while render-ing one's sense of the material body all the more dependent on it. Ford, with his "any color you want as long as it's black" philosophy, celebrated uniformity, while GM offered cosmetic differences as evidence of uniqueness. Both positions ultimately located identity—both human and automotive—in the body. And that body was hardly natural; rather, it conformed to assembly line standards.

The ideology of the assembly line held powerful sway in the modernist imagination, in which the forces of Fordism and Taylorism, as Martha Banta puts it, were "producing a nation whose notion of wholeness was inspired not by Emerson's man redeemed from the ruins but by the Model T" (277). Men were to become standardized and interchangeable, valued less for their individuality than for their efficiency and conformity. The perception of the human body could not help but be shaped by this growing dependence on machinery. Armstrong notes, "Modernity, then, brings both a fragmentation and augmentation of the body in relation to technology; it offers the body as lack, at the same time as it offers technological compensation" (3). This sounds remarkably similar to Ford's attempts to engineer better men through the use of technology; men found their identities through working on the assembly line, engaged in mass production to earn the means for mass consumption. Whatever the body lacked could be provided by purchasing the very technological products that had revealed the body's inadequacy, such as cars. As David Harvey points out, one of the basic tenets of Fordism was "his explicit recognition that mass production meant mass consumption" (125–26). Bringing the assembly line into prominence meant committing oneself to standardization and consumer culture.

But standardization has interesting implications for racial hierarchy. Ford, to his credit, employed African-American workers in more responsible positions and in greater numbers than any of his competitors—though still to less desirable jobs than white workers generally received.[5] By 1921 they constituted 10 percent of the workforce. He recruited through black churches and established a "Negro Department" (Flink, *Automobile* 126). So dedicated were his black employees that many worked throughout the 1941 strike that finally brought unionization to the Ford Motor Company.[6] The racialized body was thereby made visible in the Ford River Rouge plant at the same time as the physical body was discursively aligned to a boiler. Ford extended the worker's body to include nonwhite bodies and simultaneously reduced the body to the level of a machine and its functioning parts.[7] Furthermore, the whole idea of standardization—of bodies and machines—makes it theoretically difficult to defend ideas of segregation and hierarchy. Thus, Fordism, whatever its disturbing implications, challenged the notion of white supremacy. If all was interchangeable, how could anyone be any better than anyone else? As Ford himself put it, though probably not with this issue in mind: "Machinery is accomplishing in the world what man has failed to do by preaching, propaganda, or the written word. The airplane and radio know no boundary. They pass over the dotted lines on the map without heed or hindrance. They are binding the world together in a way no other systems can" (*Philosophy* 18). So, while Barney Oldfield may have saved the white race and the machine may have taken up the white man's burden, automotive technology was also erasing and rearranging such boundaries.

As Michael Trask has pointed out however, to accept assembly line ideology too easily "belies the violent struggles its advocates had to wage on its behalf" (20). He argues that "the period's various technologies of standardization and mechanization . . . scarcely conduce to the cultivation of routine or standardized *subjects*." Rather, "the whole notion of social placement becomes compromised by the ease with which modern culture allows persons to move in and out of spaces" (Trask 19). In *Babbitt* Lewis's characters challenge the assumption that standardization necessarily negated individuality. Even the "radical lawyer" Seneca Doane defends standardization: "Standardization is excellent, *per se*. When I buy an Ingersoll watch or a Ford, I get a better tool for less money, and I know precisely what I'm getting, and that leaves me more time and energy to be individual in" (Lewis 85). Modern writers were well aware of the multiple implications of Fordism, some supporting it and others readily perceiving its failure to produce either egalitarianism or standardization. American industry was also well aware that theoretical possibilities and practical application do not necessarily match. It may have been intellectually dishonest to continue segregation in the age of assembly line mentality, but both private and corporate America had no problems doing so. A 1921 ad for Aunt Jemima pancake flour makes that all too clear (Figure 5).[8] In it the "young engines," those with the potential horsepower, are white. The placement of the figures suggests that the little white boy facing us will exert his energy to do great things; the little white girl will sew together her jolly Aunt Jemima doll. And Aunt Jemima herself—a corporate image rather than a human being—is doubly displaced by the caricature on the box and only provides the sustenance for the body, which is, of course, white. She will not be powering or selling any machines, merely continuing to serve white bodies. White boys will grow up to master technology, white girls will continue their domestic tasks, black women will feed them, and black men are entirely left out. Technology, the ad suggests, constitutes the future and will remain the province of white men.

But Quaker Oats, the maker of Aunt Jemima pancake flour, did not have the last word on the complicated interconnection between technology, race, and gender. African-American men certainly had access to technological production in the Ford factories, and the automobile was used by people such as Jack Johnson in ways that challenged, unsettled, and demanded the reconfiguration of white privilege. Some of the most searching and perceptive authors of the era, particularly William Faulkner,

Fig. 5. Aunt Jemima Pancake Flour advertisement: "Fine Fuel for Young Engines!" (*Ladies' Home Journal*, 1921).

explored these contradictions at great length. The concerns over human identity in the machine age struck a chord in any reader of modern literature. The automobile age did not initiate the machine age, but modernist technology was producing machinery with a much greater influence on personal lives than the great factory industrialization of the nineteenth century. The telephone, telegraph, and radio all brought the world of the machine into the home. And the automobile brought the individual out into the world. While Henry Ford was facilitating this automotive victory, writers were questioning the role of men in an increasingly technological age.

One thinks of Eugene O'Neill's Yank in *The Hairy Ape* as a man reduced to a pre-evolutionary state through running the engine room on an ocean liner.

Faulkner's work provides possibly the best example of a writer who explored the tensions of the modern machine age. Although not known primarily as an industrial novelist, in his fiction he revealed the encroaching industrialization of the South, from the remark that Doane's Mill, Lena Grove's hometown in *Light in August,* will soon be abandoned once the lumber industry has denuded the timber (the same fate that overtakes Major de Spain's hunting camp in *Go Down, Moses*) to the growing realization in his late novels that, as Gavin Stevens puts it in *Intruder in the Dust,* "The American really loves nothing but his automobile: not his wife his child nor country nor even his bank-account . . . . but his motorcar" (233). If even Faulkner, a writer who grounded human identity firmly in the material world of the South and the body, saw the car as central to the human condition, it suggests that looking more closely at the role of the automobile is crucial to a fuller understanding of the modernist period. I begin with Faulkner's presentation of automobility and its link to masculinity then look to a range of modern white women writers and African-American writers to gauge the role of the car in unsettling race and gender. The pervasiveness of car culture in the literature indicates the extent to which the mass culture of automobility shaped the literary imagination, most famously in F. Scott Fitzgerald's 1925 novel *The Great Gatsby,* in which the ostentatious "circus wagon" lingers in the reader's memory as the emblem of all that Gatsby stands for. There are numerous other instances, maybe somewhat less striking, but their numbers reveal how many modern authors looked to the car as a material site of vast symbolic significance.

Faulkner, Men, and Cars

While clearly fascinated by the machine age, Faulkner's work also explores the tension between mass production and human identity. Despite the alleged isolation of the South, the car made major inroads into Southern culture.[9] A sardonic "New Psalm" submitted to Faulkner's hometown newspaper, the *Oxford Eagle,* in 1925 by an R. C. Bailey makes this clear: "The Ford is my auto; I shall not want. It maketh me lie down in muddy roads; it leadeth me into trouble; draweth on my purse. I go into the path of debt for its sake. Yea, though I understood my Ford perefctly [*sic*], I fear much evil lest the radius rod or the axles might break, or it hath a blow-out in the presence of mine enemies. I anoint the tire with a patch; the radiator boileth over. Surely this thing will not follow me all the days of my life, or I shall dwell in the house of poverty forever." Faulkner was less interested in the working of the car than in the meaning of the car. He expressed some of his reservations about automobility by linking the car with troubled white male identity and its relation to blackness and femininity. Fragile masculinity is one of the primary subjects of Faulkner's novels, in which men, both young and old, seek ways of holding femininity at bay. In "Was," the opening story of *Go Down, Moses*, Uncle Buck McCaslin remarks that in their territory "ladies were so damn seldom thank God that a man could ride for days in a straight line without having to dodge a single one" (7). He's referring, of course, in this pre–Civil War setting, to riding on horseback, for cars were as yet unheard of, and men were forced to rely on horses as a premodern form of escape. The horse apparently offered some degree of freedom from women. The car, however, would prove more problematic.

Ladies and cars reflect Faulkner's uneasiness about modernity, progress, and male independence. Even while writing about the antebellum South, he recognized the extent to which the car would tap into American desires for mobility and freedom. For the McCaslins are not just connected to horses; they are also linked to cars. Farther along in *Go Down, Moses*, Lucas Beauchamp passes judgment on the McCaslin descendants based on the vehicles they own: "There was a tractor under the mule-shed which Zack Edmonds would not have allowed on the place too, and an automobile in a house built especially for it which old Cass would not even have put his foot in. But they were the old days, the old time, and better men than these" (43–44). The degeneracy of the family line and of male identity—the ancestors were "better men"—is revealed in its increasing dependence on machinery, not just for farming but for mere convenience and pleasure. Similarly, in the old time, as Uncle Buck pointed out, ladies were seldom seen.

The Golden Age, then, came to be defined by a scarcity of women (or, at least, white women, since it is unlikely that Buck and Buddy were counting black women as ladies) and a lack of cars. On one hand, cars and women seem to have had an emasculating influence on men. But cars were also perceived to enhance masculinity. As Barney Oldfield's victory made clear, the car could redeem and redefine white manhood. In fact, it could perhaps even eliminate the need for women; many hoped that new technology could render the female womb obsolete, thus granting men control of all creativity. As Seltzer observes, late-nineteenth-century realist fiction attempted to replace "female generative power with an alternative practice, at once technological and male" (28). By the modernist era that technological generative power was fully ensconced, though not as male dominated as some could have hoped. The popularity of the Tin Lizzie—both as car and as cultural icon—certainly indicates that a significant trace of feminine power lingered in the machinery. Female generative power, whether figured through maternity or machinery, was not so easily erased. Faulkner did not replace the mother with a machine. He did, however, recognize the extent to which women and the machine were interconnected, and while he may not have agreed with Buck and Buddy that the Golden Age meant no women and no cars, he acknowledged that the combination of women and cars posed a potent threat to traditional masculinity.

Faulkner may have been uneasy about the combined power of femininity and automobility, but he was even more concerned with issues of mass production and human identity. His work issues a direct challenge to Fordist standardization. For him identity did not mean interchangeability, as it did for Ford, who discouraged any type of fellowship or individuality on the job. Workers communicated by the "Ford whisper," without moving their lips. They wiped all expression from their faces in what became known as "Fordization of the face" (Flink, *Car* 87). But even Faulkner's twin characters are unique individuals. Buck and Buddy will "fight anyone who claimed he could not tell them apart," and, it is said, "any man who ever played poker once with Uncle Buddy would never mistake him again for Uncle Buck or anybody else" (*Go Down* 7). The two men may look alike, but they are far from interchangeable. In resisting the pressure to conform to a mentality of mass production, Faulkner captured the modernist tension between man and machine. Gartman argues that mass production and consumption "undermined the bonds of occupation, class, and ethnicity" and encouraged Americans "to measure their lives in technological terms" (43, 56). Standardization elided the boundaries between man and machine, but it also, according to Gartman, weakened divisions of race. Identical twins are less identical, and blacks and whites less different, than had been assumed.

The car may have taken up the white man's burden and saved the white race, but it also made clear that such salvation was becoming increasingly fractured in the age of automobility. For example, Isaac McCaslin's misguided but understandable desire to erase the past of incest and rape by abjuring his inheritance is rendered even more problematic by his outraged rejection of the visibly white but genetically African-American woman who bears Roth's child at the end of "Delta Autumn": "*Maybe in a thousand or two thousand years in America*, he thought. *But not now! Not now!* He cried, not loud, in a voice of amazement, pity, and outrage: 'You're a nigger!'" (*Go Down* 344). Neither the automobile that brings Ike to the rapidly vanishing wilderness nor the motorboat that brings the young woman to the hunting camp has altered the existence of miscegenation; past and present alike shared troubled and abusive racial relations, which nostalgia for a lost wilderness could not eradicate. The car offered no panacea for the legacy of slavery or the exploitation of the land. At times the presence of technology only illustrated how little we control because some things never change; only the mode of transportation is different. Even a body that seemed to reveal no trace of blackness, whether it be this "African-American" woman or the more famously tragic body of Joe Christmas, was ultimately defined as racially "other." Yet that "blackness," so clearly constructed by white racism and imposed upon seemingly white bodies, itself reminds us that while race could be essentially mass-produced by a culture industry invested in maintaining racial difference, race itself was destabilized in the process. Like the car, race was something that could be made. The African-American body was, in effect, brought into being by white industry.

This industry also produced cars that challenged the dominance of the white body. In *Flags in the Dust* the black servant Simon is concerned about what young Bayard's recently purchased car means to the continuation of white privilege. He's disturbed that "Sartorises come and go in a machine a gentleman of his day would have scorned and which any pauper could own and only a fool would ride in" (*Flags* 119). Simon resents any type of leveling, being too identified with the grandeur of the Sartoris family to perceive that when a pauper and a gentleman can own the same vehicle, it may mean that he and his family can claim a kind of equality with the white gentry. Indeed, this is a possibility anticipated and caricatured by Fitzgerald in *The Great Gatsby,* in which Nick Carroway notes: "As we crossed Blackwell's Island a limousine passed us, driven by a white chauffeur, in which sat three modish Negroes, two bucks and a girl. I laughed aloud as the yolks of their eyeballs rolled toward us in haughty rivalry. 'Anything can happen now that we've slid over this bridge,' I thought; 'anything at all'" (73). "Anything" may include whites driving blacks, but Fitzgerald's language—the "bucks" and the "yokes of their eyeballs"—reifies white supremacy by mocking and dehumanizing black bodies, even as it acknowledges the challenge. Black people in cars, particularly those driven by white chauffeurs, retain a ludicrous aura, thus defusing the leveling power of automobility for Fitzgerald. Faulkner's portrayal of Simon, however, subtly emphasizes that Simon's belief in white supremacy is as outmoded as his futile resistance to the automobile.

In *The Sound and the Fury* Jason Compson, who fraudulently buys a car with the thousand dollars his mother gives him to invest in Earl's store, provides an interesting case study into cars, white privilege, and masculinity. Jason perceives himself to be a shrewd businessman, constantly chaffing at losing his chance for a job in the bank. But Jason, who must be one of the few men in America to lose money in the stock market in 1928, ultimately prefers automobility to business. Clearly, there is something about the car that means more to him than pursuing a career as a part owner of a business, rather than an employee. If being deprived of the job at the bank costs him his manhood, the car, he seems to feel, may restore it, may give him the power and control he so desperately wants. It gives him the chance to reduce his dependence on his mother, to be his own man rather than a mama's boy. He needs not just a car but a particular kind of car. Although we never learn its make, we do know that it's not a Ford: "I think too much of my car; I'm not going to hammer it to pieces like it was a ford" (*Sound* 238). Jason's thousand-dollar car is roughly double the price of most 1920s era Model Ts, necessary for him in his attempt to maintain his class status: "I says my people owned slaves here when you all were running little shirt tail country stores and farming land no nigger would look at on shares" (239). Lacking now the slaves and the land, he is reduced to claiming the car as evidence of his social and racial superiority.

While cars were initially the toys of the rich, Henry Ford promoted cars as a means of class unity—anyone and everyone could own one. He was so successful, in fact, at promulgating this vision that despite his overt hostility to socialism, he became a hero in the newly formed Soviet Union, where Russians used the term *Fordize* as a synonym for *Americanize* (Flink, *Automobile* 113). But, says Gartman, cars "united classes not in reality, by narrowing the gap of economic and political power, but merely in appearance, by obscuring class differences behind a facade of mass consumption" (15). In the modern era, however, mass consumption was more than a facade; as Andreas Huyssen points out, it sustained modernity. More than anything else, Jason's desperate attachment to his car reflects his belief in consumer culture as a means of determining identity and place in the modern world. That this status symbol proves totally inadequate to maintain either his masculinity or his family's social rank reflects just how far the Compsons have fallen: trying to shore up their position through consumer goods rather than community leadership, falling out of high culture and into mass consumption.

The car not only fails to cement Jason's class status; it also undoes his masculinity, revealing his failures as a man. He fails in his pursuit of Miss Quentin and the young man in the Ford during the afternoon chase, ending up with a flat tire, and fails in his attempt to track them down after they have robbed him, ending up, even more ignominiously, forced to hire an African-American driver to get him home. Even the smell of gasoline makes him sick. Rather than restoring his masculinity, the car reinscribes its loss, his inability to drive being linked to his failures as a man. "'Maybe I can drive slow,' he said. 'Maybe I can drive slow, thinking of something else. . . .' so he thought about Lorraine. He imagined himself in bed with her, only he was just lying beside her, pleading with her to help him" (307). The car, fraudulently if not criminally obtained, literally drives home to him his failed masculinity, and he must pay a young black man to perform what he himself cannot do: drive a car. Not only has he lost his slaves; he loses his racial supremacy and masculine privilege, defeated by the machine that was supposed to ensconce him as one of the elite. The unnamed African American replaces him as driver and thus as master of automotive technology. The man may speak in a pronounced black dialect, ostensibly marking his place in the social hierarchy, but he redeems Jack Johnson's defeat by taking charge of the car. Fordism makes men, but the car can unmake them and remake them, replacing white Jason with a black driver. If the body is composed of functioning parts, Jason's body—which cannot drive—should be sent to the scrap heap.

Cars do not just offer African Americans a foothold into dominant culture and a new configuration of masculinity. They also, of course, transform women's power. As Felski notes: "The changing status of women under conditions of urbanization and industrialization further expressed itself in a metaphorical linking of women with technology and mass production. No longer placed in simple opposition to the rationalizing logic of the modern, women were now also seen to be constructed through it" (20). If women were linked to technology and constructed through the modern, then men's relation to both women and technology became increasingly problematic. Given that the car had reshaped modern masculinity, its association with women called that masculinity even further into question. Fitzgerald, of course, saw women and cars as a deadly combination. Jordan Baker, named for two cars marketed to women, the Jordan Playboy and the Baker Electric, is a "rotten driver" (63). Nick links this failure to her dishonesty, recalling her various lies and rumors of moving a ball in a golf tournament at the same time he is accusing her of bad driving. And when Daisy is at the wheel, people die. Women in *The Great Gatsby* are safe neither in, nor from, cars. Mabel's horrific death leaves her exposed to the gaze of the crowds, with "her left breast . . . swinging loose like a flap" (145). Fitzgerald's brutal description not only details the loss of her "tremendous vitality"; it also emphasizes her female body. In this novel automobiles may make a mockery of African Americans, but they turn women into dead meat. If women are constructed through automobility in Fitzgerald's world, one sees scant hope for the future of humankind.

Faulkner, however, has a more nuanced perspective. In *Flags in the Dust* Simon laments the neglected horses and carriage, the emblems of gentility, which have given way to the automobile. Interestingly, it is the fact of the Sartoris men driving the car that disturbs him most: "It didn't make much difference what women rode in, their menfolks permitting of course. They only showed off a gentleman's equipage anyhow; they were but the barometers of a gentleman's establishment, the glass of his gentility; horses themselves knew that" (*Flags* 121). Horses may know that women are merely the "barometers of a gentleman's establishment," but cars do not. When cars replace horses, women may cease to show off a gentleman's equipage and become drivers in their own right. And when women take the wheel, the results are far less apocalyptic than in *Gatsby*.

As young Bayard careens through town on a stallion, he passes Narcissa in her car. His glimpse of her reflects all the complications of the connection between women and the automobile. Her image "seemed to have some relation to the instant itself as it culminated in crashing blackness; at the same time it seemed, for all its aloofness, to be a part of the whirling ensuing chaos which now enveloped him; a part of it, yet bringing into the vortex a sort of constant coolness like a faint, shady breeze" (*Flags* 143). Both aloof from and an integral element of the "whirling chaos," even the serene Narcissa becomes inextricably imbricated with the technology of speed and auto-mobility. The car thus offers no escape from women, and using the car to assert one's masculinity guarantees failure, if one defines masculinity as separation from femi-ninity. The stallion, a classic image of masculinity and male sexuality, is eclipsed by a woman in a car. Traditional masculinity finds itself as precariously balanced as young Bayard on the horse; appropriately, the horse immediately throws him, subjecting him to Narcissa's smothering nursing care.

In these fictional works the automobile reflects the extent to which masculinity is intertwined not only with technology but also with both blackness and femininity. At a time when race and gender were being reconfigured from a variety of angles—from women's suffrage to the Harlem Renaissance, from the "new" women to the Great Mi-gration of African Americans to northern industrial centers, where jobs in the auto industry beckoned—the car played a significant role in destabilizing and restabilizing social hierarchy. Faulkner, the greatest novelist of the modern era, captured brilliantly the challenges the car posed to human identity, particularly to masculine identity. Modernist women writers, however, presented a somewhat different picture. Just as men perceived that the car might weaken white male privilege, women recognized the same possibilities.

Modern Women Writers and Car Culture

Edith Wharton was not the only early-twentieth-century woman writer fascinated by the automobile. Many others shared her passion. Gertrude Stein, Rose Wilder Lane, and Kay Boyle all focus considerable attention on the motor car. Indeed, if we accept Andreas Huyssen's premise that modernism "constituted itself through a conscious strategy of exclusion, an anxiety of contamination by its other: an increasingly con-suming and engulfing mass culture," nothing epitomized mass culture more thor-oughly than the automobile (vii). And mass culture, as both Huyssen and Felski point out, was increasingly associated with women.[10] But the association between women and the car that proved so unsettling to writers such as Faulkner and Fitzgerald found a different twist in women writers. While it certainly did not equalize gender and race relations, the car opened the door to rethinking the very boundaries of race and gen-der by driving a wedge between identity and the physical body. With the exception of Boyle, these white women writers did not generally explore the connection to race that proves so fascinating—and disturbing—in Faulkner's work.[11] They did, however, investigate the link between the car and the body, but more with an eye to its influence on gender rather than race. The female eye is a markedly less anxious one. In the works of these women writers the automobile functions more like an extension of the self than a challenge to the self, causing us to rethink assumptions about the meaning of women's association with technology in modernity.

For Stein, as for Wharton, the car was one of the miracles of the twentieth century. As she puts it in *The Autobiography of Alice B. Toklas,* "Gertrude Stein always says that she only has two real distractions, pictures and automobiles" (210).[12] With these two hobbies linked as equals, Stein reveals her utter lack of concern for Huyssen's Great Divide; high art and mass culture equally engage her. Unlike Wharton, who was chauffeured, Stein drove, though only forward. Her partner, Alice B. Toklas, later noted: "She knew how to do everything but go in reverse. She said she would be like the French Army, never have to do such a thing" (57). This whimsical relation to the car marks a sharp contrast to the heavy symbolic meaning that resonates throughout Faulkner's work. Photos show Stein as driver, as opposed to Wharton, who was photographed as a passenger. Stein's comfort with the car comes across in the confidence with which she claims her place in the driver's seat (Figure 6). The car serves as a member of the Stein family; indeed, her first automobile, a Ford, is called "Auntie" after Stein's Aunt Pauline, "who always behaved admirably in emergencies and behaved fairly well most times if she was properly flattered" (Stein 172). This intertwining of human identity with machinery lacks the anxiety that theorists of modernism have suggested it should carry. Rather, Stein's kinship with the car and her gendering of it reinforced women's comfort with automobility.

Driving for the American Fund for the French Wounded during World War I, Stein may have ventured into a man's world, but she never appears to have perceived the situation as a challenge to gender identity; if anything, it merely served as a challenge to her driving ability, particularly her resistance to driving in reverse. In fact, she excelled at getting various men to help her with the car: "whenever there was a soldier or a chauffeur or any kind of a man anywhere, she never did anything for herself, neither changing a tyre, cranking the car or repairing it." But rather than seeing this as an indication of female frailty, Stein identifies it as a democratic ideal: "You must have deep down as the deepest thing in you a sense of equality. Then anybody will do anything for you" (174). By transforming what some would call male chivalry and others male chauvinism into an act reflecting equality, Stein refused the stereotype of the helpless woman driver and thus granted women equal status within car culture. Unlike Jason Compson's inability to deal with his car, which comes across as a failure of masculinity, Stein refused to allow the car to shape her sense of self. Auntie is a mem-

Fig. 6. Gertrude Stein and Alice B. Toklas in car. Yale Collection of American Literature, Beinecke Rare Book and Manuscript Library. Reprinted by permission of the Estate of Gertrude Stein, Stanford Gann Jr., literary executor.

ber of the family, a being whom one relates to, not an entity that asserts psychological control.

Following the demise of Auntie, Stein and Toklas acquire a new Ford, this one named Godiva "because she had come naked into the world" (191). While not literally identified as a family member, Godiva certainly appears to have a mind of her own, continuing on the motif of the car as treasured companion with whom one interrelates (Figure 7): "Back in Godiva and on the road again it was obvious that somewhere we had made a wrong turning. Was Godiva or Gertrude Stein at fault? In the discussion that followed we came to no conclusion" (Toklas 84). In their willingness to see woman and car as equals, Stein and Toklas accepted the link that so many male modernists found so troubling, fully acceding to the technological encroachment into human identity. Rather than reducing femininity, however, it increased it, adding more "women" to the population and extending human conversation to machinery. Leigh Gilmore has observed that Stein recast autobiography by relocating "the meaning of gendered identity within a contradictory code of lesbian (self-) representation" (57). By integrating the female car into a lesbian household, Stein challenged the heteronormativity of the family and of automobile culture. Early observers of automobility, as noted in Chapter One, expressed concern about women controlling a machine cul-

Fig. 7. Gertrude Stein in "Godiva." Yale Collection of American Literature, Beinecke Rare Book and Manuscript Library. Reprinted by permission of the Estate of Gertrude Stein, Stanford Gann Jr., literary executor.

turally gendered as female. Stein accepted this role as a matter of course, seeing in women's relation to the car an extension of lesbian identity.

Stein's reassessment of the implications of automobility went beyond her acceptance of the car as just another woman. For her the car provided not only a broader circle of social relations but creative inspiration by providing a work site. "She was particularly fond in these days of working in the automobile while it stood in the crowded streets" and found herself "much influenced by the sound of the streets and the movement of the automobiles" (Stein 206). More than virtually any other modernist, her prose takes on the tone of traffic, with its pulsating rhythm: "One cold dark afternoon she went out to sit with her ford car and while she sat on the steps of another battered ford watching her own being taken to pieces and put together again, she began to write. She stayed there several hours and when she came back chilled, with the ford repaired, she had written the whole of Composition as Explanation" (233). Here we see a variation on Banta's concern that Fordism was producing a nation inspired not by Emerson but by the Model T. When under the observation of Gertrude Stein, the Model T could offer considerable inspiration for innovative and experimental writing.

Interestingly, the Model T seems to have inspired a woman's tradition, aligned both with the mass culture of the automobile and with the high modernism of Stein's dense and difficult prose. It is also thus aligned with lesbian writing. Elizabeth Meese suggests that the "lesbian writer seeks to intervene in language, reinvent, or better, rework its texture, to produce an exploratory language through which we can find ourselves as subject and (of) desire" (14). By using the car to help produce this lesbian discourse, Stein both refuted women's exclusion from the age of automobility and challenged automobility as masculine and patriarchal. "Confronted with censorship, stereotyping, and devaluation of lesbian sexuality," writes Corrine Blackmer, "Stein responded by subverting the foundations of a linguistic and symbolic system that allied adult authority and propriety with male heterosexuality" (224). The symbolism of the car no longer supported heteronormativity, as Stein deployed the automobile in a way that destabilized both gender identity and any notion of technology and mass culture being opposed to art. The Great Divide, set forth so persuasively by Huyssen, was breached by a woman in a car.[13]

Stein may stand as one of the most radical purveyors of automobility, but even women of far more conservative political views shared her glee at watching the car defy the masculine order. In 1926 journalist Rose Wilder Lane, the daughter of Laura Ingalls Wilder, the author of the *Little House* books, undertook a car journey from Paris to Albania with her friend Helen Dore Boyleston. The travel journal, *Travels with Zenobia*, kept jointly by Lane and Boyleston, recounts their adventures and the responses by passersby to the sight of women drivers. Lane and Boyleston shared Stein's vision of the car as just another woman, though they lacked her commitment to viewing it as an instrument of democratic thought. And while they certainly used the car to showcase their ability as drivers and mechanics—far more than Stein, who sought to get men to do the work—their journal also reveals and produces greater awareness of the car's limitations in equalizing gender. They drove a new Ford named Zenobia, and the car apparently garnered as much attention as the fact that it was occupied by women: Lane, Boyleston (called "Troub," short for "Troubles"), and their servant, Yvonne. "Everywhere we go Zenobia makes a terrific sensation and so do we," they wrote. "Every group of people we pass nearly loses its eyes out, and sometimes they shout, 'Le nouveau modèle Ford!' and sometimes, 'Voilà trois femmes seules!' (Lookee! Three women alone!)" (34). As in Stein, one sees the slippage between woman and car as exciting rather than frightening; the combination of the latest Ford and women drivers heightens the spectacle but does not seem to challenge human identity. The women's presence is remarked upon, but they evoke little, if any, hostility from onlookers. In fact, the appeal of technology creates a kind of buffer zone in which women merely share billing with a new car rather than constituting the entire spectacle; their roles as drivers both displays and protects them.

But they also are clearly linked to imperialist travelers, out to conquer the "natives." Indeed, their often disparaging remarks about both the French and Italians— though Lane is full of praise for the improvements Mussolini had wrought in Italy, as the "whole country simply surges with hope and pride" (55)—situate them amid a privileged tradition of travelers. They may only drive a Ford, but it is the latest Ford. And as Lane boasts, upon approaching a fancy hotel: "Zenobia hardly dared enter its great gates at first. But she is a Ford of spirit, and after the first hesitation dashed haughtily between the towering pillars and saluting warders, and rolled along the curving driveway with an air that would be the envy of a Rolls Royce. In fact, she passed a Rolls Royce with exactly the right air of assured and courteous indifference" (41). Given the strong anthropomorphism associated with the car, it is a short step from the haughtiness of car to the haughtiness of its occupants. This construction inverts Stein's (and Ford's) claim that the automobile could serve to level privilege and spread democracy; here American assurance trumped automotive populism, reminding us that in this era cars did not just convey masculine privilege; they also asserted American privilege.

The car thus conveyed imperialist entitlement; despite the middlebrow celebration of a Ford over a Rolls, Zenobia and her passengers enjoyed the luxuries of the estate. The place may have been expensive, but Lane tossed it off by claiming they could always "cable" for more. Yet she follows up that comment with an even more interesting one: "We'll enjoy it and we don't care. We can always make some more money" (41). On one hand, her remark reflects the privileged position that allowed these women the means for the trip. Indeed, the car they drove is proof of that. While they complained of the prices, they could always cable for money. But they could also "always" make money. Lane and Boyleston inhabit a middle ground between Jason Compson, who seeks to uphold his class status through his car, and Gertrude Stein, who insists that the car is an instrument of democracy. Their confidence in their capacity as wage earners marks their privilege while at the same time revealing how little gender seems to matter. With the car to do the heavy labor in moving and a profession (journalism) to sustain them, they might as well be men.

But the trip also reinforces gender distinctions. After arriving in Albania, they decide on a short excursion, and, "getting lazy," they employ a young chauffeur (103). The car acts up, and Troub retakes the wheel, much to the boy's dismay. He "all but howled. We told him in three languages that Troub knew more about driving a car than he'd ever dreamed of. We put him in back. And he kept leaning forward, draping himself around her shoulders like a feather boa, and trying to tell her to hang on to the wheel, and not put—Oh God, oh God!—her foot on the foot throttle—which of course she did!—and that the gasoline was the right-hand throttle, and that the emergency brake was the emergency brake" (106). It is their gender, of course, that inspires the young man's performance, reminding the women that they are not supposed to know what they are doing. Ultimately, it is their gender that defeats the young Albanian chauffeur; he ends up "disconsolate, and mourning to himself and to Allah in low, heart-stricken tones" (108). The combination of driving, mechanical, and linguistic prowess marks them as superior, economically able to hire help yet literally able to do the job better and, in so doing, to assert female knowledge and agency. Yet they also bear out Rosemary Marangoly George's admonition that colonialist women may have enjoyed "greater political participation and greater personal authority than feminists and women modernists of the time in England. The memsahib was a British citizen long before England's laws caught up with her" (62). Certainly, in the journal's narrative Lane relishes her victories over the car and the natives.

There is no doubt that Lane and Boyleston exercised considerable authority and enjoyed the kind of white privilege that so many modern men sought. But despite their imperialist privileges, they could not leave gender behind. The automobile may have revised traditional gender roles, but it did not erase them. While Lane and Boyleston acted with the assurance of men, their sex relegated them to performing this expertise over and over again, thus effectively denying them full masculine privilege. Even such accomplished women could never have their automotive skills taken for granted. The triumph expressed in the narration of the episode serves to highlight that they are the exceptions that prove the rule of women's presumed incompetence in automotive matters—and thus their subordinate position remains fundamentally unchallenged. Despite their capability, they reflect more of Faulkner's uneasiness regarding the car's link to gender than Stein's comfort with it.

Both Stein and Lane explored the car's relation to female identity and the limits that automotive power granted women. Kay Boyle, like Faulkner, focused more on the car's association with the body. Boyle's recently discovered first novel, *Process*, offers a fascinating glimpse of how a woman uses the car to express sexuality and to explore racial difference. Kerith, the young female protagonist, seeks to position herself as both mind and body, a task facilitated by the automobile. This experimental female bildungsroman—Boyle's answer, Sandra Spanier notes, to Joyce's *Portrait of the Artist*—follows a young woman's attempts to find a balance between the material and intellectual worlds, between art and the body, a process in which the car plays an integral role (xviii).[14] The novel begins with Kerith driving along the river: "The motor sang with her, jerked words into air that was heavy and shaped with smoke" (1). The affinity between Kerith and the car follows the same patterns set up in both Stein and Lane: to merge woman and car is to enhance woman and, in this case, to create art as it "jerks" the words into the air. Just as Stein is inspired by the Ford, Kerith sings a duet with her car. In this novel the car creates a link between the female body and art.

It is also strongly tied to sexuality. In an erotically charged scene near the end of the book, Kerith speeds toward a railway crossing, just as a train arrives. "The car was descending with such unbroken speed, as though it could never be stopped, and sensitive to her fingers, with a sardonic evil response, in perfect unity with her and in perfect opposition to any other command" (73). She stops the car just in time to avoid a fatal collision, "quiet and thrilled with this, knowing that her body had scarcely moved to stop them, and that it was the bond and belief and pure subjection between herself and the car that had willed them into stopping. She might have sat there, she thought, and not lifted a finger, and yet the car would have stopped because it was in perfect response to her, and she was the attractile body drawing it to her direction and will" (73–74). The car functions as both a physical and mental extension of her body, a "pure subjection between herself and the car." This remarkable passage offers a fascinating possibility for rethinking the connections between women, technology, and modernism. Rather than separating her from the body, the car reinforces the power of the female body. For it's important to note that under the hands of other drivers, cars do not necessarily stop when willed. Early in the text Kerith comes upon the aftermath of an accident in which a motorcyclist has been killed by a car. Her connection to her vehicle, then, marks her as different and redeems the automobile from the fatal associations Fitzgerald presents. Women and cars create art and preserve life.

Kerith also seeks the "natural," and she uses the car to bring her in contact with nature. Tellingly, she finds blackness—as well as her own whiteness. On one of her trips to the country she and her suitor, Brodsky, encounter an African-American man from whom they acquire potatoes and alcohol. Kerith watches the man, his "nipples ringed vibrant and blue as fungus," and feels "the white stretch of her skin like a sterility upon her" (18). The car enhances her body, but interaction with a black man seems to diminish it into sterility. Although the symbolism of the fertile black body comes across as rather heavy-handed, Boyle subtly reminds us of this contrived association as Kerith tries to isolate the man in nature, asking him, "You *are* free, aren't you?" His response, that he pays taxes like everyone else, resists the racializing of the mind-body split. There is no purity of the natural; even African Americans in the wilderness pay their taxes, just as women, so often associated with the body, can become one with the car. It is striking that an encounter with blackness reminds Kerith of the "sterility" of her whiteness, but her connection to her car grounds her in an overtly sexualized body. Thus, Kerith's quest for the physical lies in automobility, not in a return to a stereotypical naturalized black body. With this move Boyle challenges discourse surrounding the Jackson/Oldfield race, a discourse that pits white technology against the black body. Here technology returns one to the body, to the kind of natural element also evoked by the unnamed African American. Kerith serves as a link connecting female sexuality, the physical body, and race, illustrating a different picture of modernism and technology, one that sees bridges rather than divides.

African-American Writers and Automobility

As the Oldfield/Johnson race made clear, the combination of African Americans in cars proved disturbing to white America. The motor car's ability to showcase black bodies derailed the association between whiteness, progress, and technology that so many saw embodied in the automobile. Oldfield's victory, after all, was supposed to put Johnson in his "place," reminding African Americans of their marginality in automobile culture and mainstream American culture. But as Boyle and Faulkner recognize, neither Johnson nor the African-American population accepted this exclusion from the automobile. Nor did African-American writers. Jean Toomer, Zora Neale Hurston, Richard Wright, and Ann Petry all document the role of the car in the black community. It may find scant attention in the early part of the century, but directly after World War II, with the auto industry back in full swing, Petry offers a fascinat-

ing and complex vision of the automobile's connection to race and gender, an awareness that grows out of modernist explorations of cars, race, men, and women.

Jean Toomer's 1923 *Cane*—a collage of short fiction, vignettes, and poetry—refers
to cars only in the second section, which focuses on northern urban life. The mention
is brief but is prominently displayed in the poem beginning the first piece in the section, "Seventh Street":

> Money burns the pocket, pocket hurts,
> Bootleggers in silken shirts,
> Ballooned, zooming Cadillacs,
> Whizzing, whizzing down the street-car tracks.   (39)

This poem is followed by a single paragraph describing Seventh Street as a "bastard
of Prohibition and the War." Toomer then repeats the poem to close out the vignette.
The brevity of the entire piece intensifies the effect of the "zooming Cadillacs," which
Toomer associates with illicit activity but also portrays as "whizzing" past the street-
car tracks, suggesting that the car is fast outstripping mass transit, the transportation
of the past. He paints the moment by insisting on the Cadillac rather than the Model
T, highlighting the excitement, glamour, and danger of the age, a time when money
"burns the pocket" and blacks can drive Cadillacs. Toomer presents the car as a rather
shaky avenue into the high life, a way for African Americans to participate in the glitz
and danger of the jazz age. The car helps to shape the scene with its promise yet also
warns of the dangers of the time when "Black reddish blood" flows freely (39). The
car, rather than providing significant power, may simply deliver African Americans to
death in a very turbulent era.

Zora Neale Hurston, on the contrary, places the car in a rural setting in her first
novel, *Jonah's Gourd Vine* (1934), in which, again, a Cadillac leads to death. When
Sally, John Pearson's second wife, gives him a car as an anniversary present, he feels
the gift is too extravagant; she shouldn't have gotten him a car, let alone a Cadillac:
"But uh Chevrolet would uh done me. You didn't hafta go buy no Cadillac" (193).
Sally, however, has internalized the cultural messages regarding the status of automobile culture. Like Jason Compson, she wants only the fanciest car, albeit for a more
selfless reason: to express her love rather than her social position, though she does
seem to gain social satisfaction in being able to afford the gesture. It is clearly important that this African-American woman is able to purchase a Cadillac. In this novel,
as in *Their Eyes Were Watching God*, Hurston makes clear that not all black women are
poverty-stricken. The car serves as a challenge to prevailing assumptions about the
financial position of African Americans, particularly in the rural South. The automobile is by no means financially off-limits.

But while John Pearson may have a wife who can afford to give him a car, that car does not necessarily empower him, for it sets him up as a prize to other women who "gone crazy 'bout dat car" (196). John succumbs to temptation and, guilt-stricken, is killed by driving the car into a railroad crossing "but half-seeing the railroad from looking inward" (200). The ability to "see" is critical in Hurston's work, and John's failure reflects his inability to function in the material world, whether that entails remaining faithful to his wife or driving his car without plowing into a train. The car may convey status, but it does not appear to enhance power or agency. It simply makes being a man even more difficult by delivering him into situations of temptation. As Toomer also illustrates, cars can be dangerous, especially to the weak.

The power of the car finds more significant expression in Richard Wright's 1940 novel, *Native Son*. When Bigger Thomas reports for his job as chauffeur to the white Dalton family, he hopes the car "would be a Packard, or a Lincoln, or a Rolls Royce." He looks forward to being alone in the car, when "he would burn up the pavement; he would make those tires smoke!" (61). Although he is slightly disappointed when it turns out to be a Buick, he nonetheless feels the power the car offers him: "He had a keen sense of power when driving; the feel of a car added something to him. He loved to press his foot against a pedal and sail along, watching others stand still, seeing the asphalt road unwind under him" (63). Bigger sees his powerlessness reversed by a car, the more expensive the better. He doesn't need to own it; to drive it is sufficient. The car provides not just power but "something" less defined, something that becomes a part of him. It enhances his body and thus his blackness. Bigger, who has also expressed the desire to fly planes, recognizes that control over movement and technology conveys substance and belonging. Denied that hope, he sees no alternative other than violence because his world is already a prison: "They do things and we can't. It's just like living in jail" (23).

But the promise of the car fails Bigger. Only briefly does he feel in control; only in his imagination can he make the tires smoke. As he comes to learn, being a black chauffeur is not quite the same as being a driver. He worries about endangering his job when Mary demands that he drive her to her boyfriend's rather than to the university, the destination she had announced to her father. And he is displaced even as chauffeur when Jan takes the wheel and insists on driving to the South Side, the Black Belt, to find a place "where colored people eat." Once he relinquishes the driver's seat, Bigger feels crammed in with Jan and Mary on either side of him, afraid to move because "his moving would have called attention to himself and his black body" (69). While he initially feels the car reinforce his body, it quickly becomes a trap for that body, reminding him of its blackness—and its powerlessness. The car may provide him with figurative power, but it cannot change the material conditions of his life. In essence he is reduced to a eunuch, asked to drive Jan and Mary while they make out and talk in the backseat about wanting to know Negroes who have "so much emotion" (76). They may want to *see* Bigger as a man, but they treat him as a chauffeur— and one without the power to control the car in any way except to keep it on the road. Denied recognition as a man, Bigger moves to violence to create a new life for himself. The car's promise of masculine power falls short when a black man drives a white man's vehicle. While Jason Compson suffers the loss of manhood in relying on a black chauffeur, we see in Wright that the black chauffeur is not necessarily the recipient of what Jason loses. Yet the car also offers Bigger the only means—other than murder— of feeling in charge of his own life and his own destiny. It may be fleeting, and it certainly does not convey the power of murder: "He had murdered and created a new life for himself" (101). But it does linger as possibility, as an alternative to violence, as a place where an African-American man could potentially be a man.

It is a woman writer, however, who most clearly articulates the links between race, masculinity, and cars. Ann Petry's 1946 novel, *The Street*, reflects some of the modernist questioning about race, gender, and automobility yet with greater dismay at how little the car has changed things, particularly for African-American women. As Lutie Johnson rides with Boots Smith through the streets of New York and out into the country, she comes to an important understanding of what the car can offer African Americans:

> And she got the feeling that Boots Smith's relationship to this swiftly moving car was no ordinary one. He wasn't just a black man driving a car at a pell-mell pace. He had lost all sense of time and space as the car plunged forward into the cold, white night.
>
> The act of driving the car made him feel he was a powerful being who could conquer the world. . . . It was like playing god and commanding everything within hearing to awaken and listen to him. . . . And she knew, too, that this was the reason white people turned scornfully to look at Negroes who swooped past them on the highways. 'Crazy niggers with autos' in the way they looked. Because they sensed that the black men had to roar past them, had for a brief moment to feel equal, feel superior; had to take reckless chances going around the curves, passing on hills, so that they would be better able to face a world that took pains to make them feel that they didn't belong, that they were inferior.
>
> Because in that one moment of passing a white man in a car they could feel good and the good feeling would last long enough so that they could hold their heads up the next day and the day after that. And the white people in cars hated it because . . . possibly they, too, needed to go on feeling superior. Because if they didn't, it upset the delicate balance of the world they moved in when they could see for themselves that a black man

in a ratclap car could overtake and pass them on a hill. Because if there was nothing left for them but that business of feeling superior to black people, and that was taken away even for the split second of one car going ahead of another, it left them with nothing. . . . Yes, she thought, at this moment he has forgotten he's black. At this moment and in the act of sending this car hurtling through the night, he is making up for a lot of the things that have happened to make him what he is. He is proving all kinds of things to himself.

(157–58)

Here we see the power of the automobile allowing African-American men to overtake white supremacy, "even for the split second." The car makes up for the "things" that have made Boots Smith "what he is." The wording reinforces the modernist sense that bodies are made, that Boots is constructed by forces beyond his control. The car may add "something" to Bigger Thomas, but Boots sees it as a way to play God. It helps him to face a world that denies him belonging and equality. As Gilroy points out, "For African-American populations seeking ways out of the lingering shadows of slavery, owning and using automobiles supplied one significant means to measure the distance traveled toward political freedoms and public respect" (94). Yet by focusing on the symbolic power of automobility that allows him to play God, the black body seems left behind. Boots, in fact, "forgets" that he's black, which hardly allows him to "save" the black race in the same way that Oldfield saved the white one. The car allows him, only briefly, a glimpse of white privilege; it does not increase black power. Yet though Boots may forget that he's black, it's clear that white people cannot forget it and cannot restore the "delicate balance" of racial hierarchy, at least not while on the road. As even the "bucks" in *Gatsby* reminded us, machines will carry black bodies without recognizing racial difference. The scene may not highlight the power of the African-American body as body, but it reminds us of the power of car when driven by a black body. While this does not break the power of racism, it does fracture it for a "split second."

This success, however, is gendered: Boots is the driver, the one savoring his moment of victory; Lutie Johnson is merely along for the ride, the spectator to male driving prowess and recklessness. She realizes that to him she is simply a "pick-up girl" and only manages to stave off rape by reminding him that he must get back to the nightclub for his band's gig (161). Thus, Petry documents the power of automobility to convey masculine privilege while acknowledging its significant limitations. While Lutie fears the influence of the street over her son, her real vulnerability lies in her black female body, the object of desire to Jones, to Boots Smith, and to the nightclub owner, Junto. She has no car to drive off in. Cars may allow such men as Boots Smith the opportunity to play God, but they merely serve to remind women of their physical vulnerability, as Lutie risks death by his reckless driving and rape by his superior physical strength. White writers such as Faulkner recognize the potential the automobile offers nonwhites, but Petry reminds us that as long as women—black and white—are subject to physical violence and poverty, no machine can remedy the situation. Only for a brief moment can Lutie even grasp the possibility of black power—and that moment occurs in a car. But until she can glimpse the possibility of female power, the car falls short of its promise of a different future. Despite the potential of undoing race and gender that modern literature experiments with, modernity ends with the hierarchy largely still in place and power still concentrated in white male hands.

"We are *not* living in a machine age," Henry Ford reminds us; "*we are living in the power age*" (*Philosophy* 40). But power, of course, is conferred by machines, displacing the human body as the central means through which human identity is located and grounded. By exaggerating the power—or lack thereof—of the man in control of the car (e.g., Jason Compson or Boots Smith), the car reminds us of the fragility of the body in automobile culture and the attendant fragility of the masculinity of the man who drives the machine. Could he still be a man without the car? And if a black man can drive as well as a white one, how does that undercut white masculinity if masculinity is now inseparable from technology? Jack Johnson's triumph was physical; his powerful body defeated all challengers. Because his triumph was limited to his bodily prowess in a new age of automobility, however, the old pattern remained in place: when nonwhite men and women began to close the gap of achievement, the rules of the game changed. Auto racing replaced boxing, once African Americans triumphed; machines replaced bodies once nonwhite bodies were admitted into physical competition. Today the sport of auto racing still seems predominately white. In the April 25, 2005, episode of the comedy program *The Daily Show*, host Jon Stewart commented that former secretary of state Colin Powell would be driving the pace car at the Indianapolis 500 and reminded the audience that Powell would thus be setting precedent for more than simply being the first former Cabinet member to perform this role. Stewart then offered a doctored—but very apt—graphic picturing a banner hanging at the racetrack with the words, "Welcome, Black Guy." Yet despite the apparent whiteness of racing, the car has opened up a dialectic between the fixity of the machine age and its power to begin unfixing what used to be thought of as firm principles of gendered identity and racial hierarchy.

Women writers and women drivers change the terms of the equation, claiming automotive power as integral to the female body without necessarily threatening the integrity of that body. If mass culture, as both Huyssen and Felski argue, is associated with women, then the triumph of automobility begins to challenge male hegemony;

men now have to share the roads. Although we see little acknowledgment of issues of race in modern white women's texts, Petry's harsh depiction of the car's potential and its failed promise issues a strong challenge to automotive culture, white hegemony, and future women writers. At the least the car serves as a kind of bridge between bodies and technology, ungendering and unracing power and suggesting that modernity is fully situated within the culture of automobility, a position that begins to redefine human identity as automotive identity, and vice versa.

CHAPTER THREE

My Mother the Car?

Auto Bodies and Maternity

In September 1965 one of the strangest sitcoms ever debuted on NBC: *My Mother the Car*. It had a blessedly short life, lasting only a year. Dave Crabtree, played by Jerry Van Dyke, goes to a used car lot in search of a cheap family car. Finding himself drawn to a 1928 Porter (a car that has never existed, assembled for the show from parts of various antique cars), he gets behind the wheel only to have the car begin to talk to him. Voiced by Ann Sothern, the vehicle informs him that she is the reincarnation of his dead mother. He buys the car and spends the series guarding "mother" against various pitfalls, including an antique car dealer who keeps trying to commandeer the vehicle, and it remained slightly unclear whether discovering the reincarnation of one's mother constituted a blessing or a curse. This utterly preposterous premise may seem more than a bit far-fetched, but the previous year Van Dyke had turned down the title role for the series *Gilligan's Island*, which was to run on CBS from 1964 to 1967, claiming the script was too silly to work. That he signed on for *My Mother the Car* indicates perhaps both his lack of judgment and his belief in the fascination of a mother/car combination. Indeed, when we consider some of the other popular shows of the day, this one doesn't seem so absurd after all. The much more successful *Mr. Ed* had just finished a four-year run; apparently, talking male horses are more entertaining than talking maternal cars. Overlapping with *My Mother the Car* was *My Favorite Martian*, with a three-year run from 1963 to 1966. Also debuting in 1965 were *Lost in Space*, which ran until 1968, and *I Dream of Jeannie*, running through 1970. The most famous of them all, *Star Trek*, began a three-year run in 1966.

This canon of mid-1960s television illustrates some interesting perspectives on cultural perceptions of the relationship between women and technology, issues that also find voice in contemporary women's literature and feminist theory. While Americans clearly were fascinated by space exploration and technology, welcoming Martians and space aliens onto their TV screens, they seemed less taken with talking cars.

It may be that cars appeared passé, less exciting than the new space program, which represented the technology of the future. After all, *My Mother the Car* participates in a kind of nostalgia with its "antique" car and reclaiming of the past: regaining one's dead mother. But space excitement hardly accounts for the popularity of a show in which a man—even if he was an astronaut—owned a genie who could work miracles for him and then get corked up in a bottle whenever she became troublesome. Obviously, gender intersects in interesting ways: women should be contained in sturdy receptacles, not free to reincarnate themselves in machinery. A woman with supernatural powers was not a threat as long as there was a man there to control her; the somewhat more feminist show, *Bewitched* (1964–72), nevertheless had a bumbling husband who kept trying to insist that his wife refrain from exercising her witchery. But despite the fascination with technology—or maybe because of it—the American public wanted no part of a show that insisted that a woman and a car could be one and the same.

In fact, however, American culture has been making that assumption for a century. Early terms for car parts came from women's clothing: the bonnet to hide the engine and the skirts to encircle the machinery. The massively popular Model T was nicknamed the Tin Lizzie and Henry's "lady." The second half of the century proved no less fascinated with the equation of woman and car. Stephen King's novel *Christine* (1983) plays upon the fear of a car taking on the characteristics of a jealous woman. Women and cars are the joint subjects of hundreds of popular songs; Chuck Berry and the Beach Boys have produced some of the most memorable. But to associate women and cars is one thing; to associate mothers and cars is something else. The continued belief in the sanctity of motherhood renders such a connection particularly problematic. Mothers are natural; cars are mechanical. Mothers create life; cars take it away. Thus, when mothers and cars intersect, both automobility and maternity get reshaped. We may not have tuned in to watch *My Mother the Car* in the 1960s, but the mother becomes the car in Louise Erdrich's 1984 novel *Love Medicine*, with her 1994 novel *The Bingo Palace* continuing the trope. The slippage between mother and car is also worked out in less overt fashion in Flannery O'Connor's "The Life You Save May Be Your Own," Joan Didion's *Play It as It Lays*, and Toni Morrison's *Paradise*. The mother doesn't have to *be* the car in order for the connection between mother and car to destabilize maternity and gender. To a far greater extent than the modernist literature explored in the previous chapter, contemporary women's fiction plays with the intersection of woman and machine and the mother's position in this increasingly cyborg world. Now that technology has denaturalized maternity, women's literature, which has such a strong association with motherhood, explores the role that cars play in this denaturalization—and occasional re-naturalizaion—of the mother.

No aspect of female identity endures greater cultural scrutiny than maternity. A powerful belief in motherhood as woman's "highest" or "most natural" function persists, even in the face of technology that may ultimately eliminate the female body as the cornerstone of that very maternity. Women writers, too, face the pressure of how to represent mothering; the influence of nineteenth-century ideology regarding motherhood and women's literature hovers over even contemporary women's fiction. In fact, twentieth-century writers confront possibly even greater cultural scripting, from academic discourse to the popular media to advertising. Given the power of technology and media to reshape motherhood, we must first examine the culture of maternity, and the automobile's role in it, to which these writers responded. No one reading—or writing—contemporary fiction can afford to ignore the powerful messages of what it means to be a mother. Women's writing about mothers and cars is necessarily colored by both the ideological sway of technology and the pervasive voices of the automobile industry.

Creating Automotive Maternity:
Gender, Technology, and Advertising

Technology, of course, has far-reaching implications not just for maternity but for gender itself. In *Technologies of Gender* Teresa de Lauretis, drawing on Foucault's term *technology of sex* postulates a "technology of gender" in which gender "is the product of various social technologies, such as cinema, and of institutionalized discourses, epistemologies, and critical practices, as well as practices of daily life" (2). In other words, gender is produced by a combination of technology, social institutions, and daily interactions. De Lauretis makes a very significant move in transforming the term *technology* into something literally associated with technology in the sense of being linked to various kinds of machinery or mechanization. Foucault, on the contrary, seems to use *apparatus, techniques,* and *technology* interchangeably. But de Lauretis's grounding of the term in a more concrete framework opens up ways of conceptualizing gender as shaped by a culture of technology—which also distinguishes her position from that of Judith Butler, though both share a belief in gender as having been produced.

De Lauretis's focus is on film, and many current theorists of technology and gender concentrate primarily on various forms of computer, media, and medical technology. But automotive technology, even more than more recent technological advances, has shaped not just American culture but also American configurations of gender. We tend to anthropomorphize our cars much more readily than our computers. The computer is still largely a tool; the car is an extension of self and an expression of identity. As such, it fits well into Donna Haraway's identification of what she terms our "cyborg" world: "By the late twentieth century . . . we are all chimeras, theorized and fabricated hybrids of machine and organism; in short, we are cyborgs" (150). She posits a "leaky distinction" between human and machine in an age in which machines, unlike those of an earlier era, are not haunted because they "have made thoroughly ambiguous the difference between natural and artificial, mind and body, self-developing and externally designed, and many other distinctions that used to apply to organisms and machines" (152). Those leaky distinctions, she goes on to say, offer significant potential for feminist analysis in "the breakdown of clean distinctions between organism and machine" and also in rendering "Man and Woman so problematic" (174, 176). If we're born of technology, rather than of woman, then what's the difference between men and women? Is maternity still a viable concept? By eliding the boundary between human and machine, automotive technology opens up a space for women to assert a kind of power—of autonomy, speed, and mobility— formerly associated with men. It also reconstructs maternity into something intricately intertwined with automobility. Does the cyborg, however, usurp the mother?

Certainly, the mother's body has become more and more a technologized object: pregnant women can be forced not only to carry an unwanted fetus to term; they can also be forced to submit to various invasive procedures to insure "the best interests" of the fetus. They are policed in public, rebuked for smoking, criticized for drinking coffee, refused alcohol, and additionally charged with child endangerment if discovered taking drugs.[1] They also carry considerable automotive responsibility. In *Unbearable Weight* Susan Bordo cites the example of a Massachusetts woman charged with vehicular homicide of her fetus when she miscarried after a car accident caused by her drunk driving (82). Mothers bear a special burden in car culture. The automobile industry offers no relief from such disciplinary action. By appealing to child safety, women are pressured into buying increasingly elaborate car restraint systems as well as the vehicles that provide the industry with its biggest profits: sports utility vehicles (SUVs) and minivans. Tom and Ray Magliozzi, who receive so many calls to *Car Talk* from and about mothers that they have released a CD entitled *Maternal Combustion: Calls about Moms and Cars,* tell a new mother who prefers sports cars to minivans to "give it up." Now that she is a mother, they tell her, "those days are gone forever."[2] Mothers need to drive "mom cars." Child-rearing sites routinely include sections on auto safety. Obviously, at some level this is not a bad situation; doing everything possible to give birth to healthy babies and protecting those children in car accidents, one of the biggest killers of children in the country, are admirable concerns. But the conjunction between increasing maternal technology and increasing emphasis on automobile safety positions the mother as a part of the technology that both reproduces and drives.

This leads to what I term "automotive maternity," a condition in which one's role as a mother relies, at least in part, on one's place in car culture—as driver and as vehicle. Automotive maternity situates mothers as both essential and cultural, a woman who drives the right car. One manifestation of automotive maternity can be seen in the so-called soccer mom, the woman who drives a minivan, transporting children and equipment to multiple activities, a woman who, apparently, spends her time serving (and driving) her kids. Or we can identify it in the mother who calls *Car Talk* complaining that her adult daughter does not seem to recognize the need for rotating the tires on her car. The Magliozzis, however, tell the caller, who wants advice on how to convince her daughter to take tire rotation seriously, to back off; she's trying to control her daughter's automotive behavior and needs to learn to let go, to let her daughter make her own decisions and mistakes regarding cars—and life (*Maternal*). Automotive maternity is partly composed of giving maternal car advice.

But automotive maternity has a darker side as well. Witness, for example, the national outrage over South Carolina mother Susan Smith's actions in October 1994: strapping her young sons into their car seats, driving the car into a lake, and leaving them to drown. The incident resonates not simply because this was a mother who killed her children; this was a mother who twisted one of the greatest safety advances—car seats for children—to do so. In June 2001 Andrea Yates, a mother in Texas who drowned her five children in the bathtub, did not receive nearly the same degree of acrimony. Admittedly, there were significant differences in their situations, one woman getting rid of her kids in order to facilitate a relationship with a new boyfriend, and the other suffering from severe postpartum depression. Furthermore, Smith went on to perpetrate a massive hoax by claiming to have been carjacked by an African-American man, thus launching an expensive manhunt as well as reinforcing racist fears. Yates, on the contrary, made no attempt to deny her actions; she simply dialed 911 and announced that she'd killed her children.

Nevertheless, the means by which Susan Smith committed her crime—the car—exaggerated the sense of her "unnatural" behavior. Her actions reveal the displacement of the mechanical into the human; neither she nor the car showed any sense of "heart." When Smith used the car to commit deliberate infanticide, a shocked public was reminded that mothers and cars can be a deadly combination, that the mother who drives may also be the mother who kills. Mothers bear the primary burden of transporting their children both in the womb and in the car, suggesting a disconcerting slippage between uterus and car. In fact, in explaining the popularity of the minivan as a mom car, Clotaire Rapaille, a psychological analyst with Chrysler, claims that "minivans . . . evoke feelings of being in the womb, and of caring for others. . . . Stand a minivan on its rear bumper and it has the silhouette of a pregnant woman in a floor-length dress" (qtd. in Bradsher 99).

Automotive maternity, then, situates women as transporters—similar to the legal and medical discourse surrounding pregnant women's bodies as receptacles for the fetus. In order to be a good mother, one needs to know about cars. And this highlights yet another aspect of automotive maternity: economics. Automobile safety does not come cheap—at least, according to the auto industry, which touts bigger and heavier vehicles as safer. As the cultural demands to be a good mother grow heavier, it becomes increasingly clear that being a good mother relies upon money. Mothering, like a car, is something that can be purchased; in fact, it can be purchased by buying a car. Such assumptions prove particularly problematic, as we will see, in the work of Toni Morrison and Louise Erdrich, who principally delineate the concerns of non-affluent communities, in which buying a car is not something to be taken lightly. Like mothering, it resonates throughout the community.

But automotive maternity is constructed even more powerfully in advertising; Roland Marchand uses Jacques Ellul's term *integration propaganda* to describe the function of ads (xviii). Because these messages "integrate" every level of our culture, they provide a valuable context for considering the literature. No business relies more heavily on ads than the automobile industry. According to Keith Bradsher, one in every seven dollars spent on advertising comes from car manufacturers, more than twice as much as the next two largest contributors combined: financial services and telecommunications (277). Consequently, car ads permeate magazines, radio, and television, bombarding us with images of the car's role in American life, images that often situate women as a part of the machinery. Although the ads of the 1960s featuring women draped across automobile hoods have largely fallen by the wayside, beautiful women still pose in front of cars, linking female sexuality with automotive power, both of which are to be mastered by men. And while the auto industry has long wanted to sell cars to women, it took automakers a surprisingly long time to develop interesting ads targeting women and to get their sales forces to stop calling women in their showrooms "honey"; as late as 1992, a study indicated that 39 percent of women trying to buy cars were subjected to "terms of endearment" (Hollreiser 15). Gloria Steinem, writing about the difficulties of finding advertising for *Ms.* magazine, notes that it was a ten-year struggle to persuade Detroit automakers to consider placing advertising in a women's magazine. "But long after figures showed a third, even a half, of many car models being bought by women, U.S. makers continued to be uncomfortable addressing women." Yet when such ads did appear, they presented a very problematic relationship between a woman and her car. As Steinem recalls, gender-neutral ads were few and far between: "*Ms.* readers were so grateful for a routine Honda ad featuring rack and pinion steering, for instance, that they sent fan mail" (20).

The message of these ads is so important because, as James B. Twitchell argues in *Adcult USA,* advertising does not sell goods; it sells meaning: "What advertising does, and how it does it, has little to do with the movement of specific goods. Like religion, which has little to do with the actual delivery of salvation in the next world but everything to do with the ordering of life in this one, commercial speech has little to do with material objects per se but everything to do with how we perceive them" (110). And, according to Twitchell, how we perceive such goods shapes how we perceive our lives: "In giving value to objects, advertising gives value to our lives" (4). Advertising also gives value to literature, Jennifer Wicke suggests, pointing out the "centrality of advertisement to modern culture, and its radical reshaping of both literature and ideological production" (2). While Wicke's study examines nineteenth- and early-twentieth-century British culture, her argument on the impact of advertising also helps to illuminate late-twentieth-century American culture. Given the power of advertising to shape not just our desires but also our lives, it is not surprising that writers would take on the embedded ideology of ads. The value that car ads give women's lives says much for the ways that women are valued—or not—in American culture. Certainly, the industry has always tried to appeal to woman as individual and woman as mother; the two Ford ads discussed in Chapter One target the businesswoman and the mother. The attempt to target female buyers has only strengthened throughout the century. Beginning with contemporary ads linking women and cars and then considering how such associations play out when the female body is also the maternal body, the following discussion maps out the cultural context that so many female novelists strained against.

A motif that had been around at least since the Pierce Arrow ads of the early century—linking classy cars to classy women—still found frequent expression in the late 1970s. Its popularity helps to explain why Steinem had such difficulty landing "standard" car ads. In 1979 *Harpers Bazaar* featured a run of ads by Cadillac and Jordan Marsh, the upscale department store. Picturing a beautifully dressed woman next to a Cadillac Seville, the copy reads: "One of the world's best cars. It expresses its excellence in every smoothly proportioned line, every beautifully crafted detail. As do the Sanford Sports separates, attuned to resort now and the north later. Softly stated in teal." The same language applies to cars and clothes; women can only appreciate the car, the subtext suggests, if it is presented as part of their wardrobe, with the same "proportioned lines" and "beautifully crafted details." Buying a car is no more complicated than buying resort wear, requiring not brains so much as taste, something

Fig. 8. Ford advertisement: "Detroit Style" (*In Style,* 2001).

women were expected to possess naturally. And even this conservative ad, with its snobbish appeal to Cadillacs and resort wear, reflects a slippage between woman and car, both well-groomed, expensive luxury items.[3]

By the turn of the twenty-first century, however, such slippage becomes far more overt, even in more "feminist"—or rather, postfeminist—ads. Toyota, advertising in the August 2000 *Better Homes and Gardens,* seems to offer an overtly feminist appeal with its copy: "I am Corolla. Hear me roar?" Evoking the 1972 Helen Reddy song often viewed as a feminist anthem—"I Am Woman"—this ad suggests that its car is powered by feminist energy. In fact, there is no need to feature an actual woman—and no woman appears in the ad; the woman has become the car. This erasure casts some doubt over the allegedly feminist message; in car culture, it suggests, women are not required. If the car functions as a cyborg, thus replacing the woman, we run the risk of eliminating women altogether. It seems as if Toyota has ingested her and displaced her voice into a car. The woman's body no longer "roars"; the auto body does, as cyborg trumps woman.

The ultimate cyborg ad, however, appeared in 2001 in *In Style* for a Ford Focus; here the woman has apparently ingested the automobile (Figure 8). It is particularly striking that here what becomes mechanical are precisely the female sexual and reproductive organs. By replacing the female body from breasts to upper thighs with automotive machinery, the ad effectively erases female difference. Women, it suggests, have no biological function; they are simply mechanical objects. While it's encouraging to see the definition of woman as baby maker challenged, this hardly seems a triumphant alternative. Or maybe the message is that women's biological function is to produce cars, a fascinating cyborg possibility. Even more disturbing is a Lexus ad appearing in *Esquire* in September 2000 portraying the woman not as a cyborg but as an even more lifeless piece of "art" (Figure 9). Picturing a naked woman apparently bound and gagged juxtaposed to a Lexus, the copy claims to distinguish between the two: "Are we the cutting edge of avant-garde. . . . Well, no." The woman is apparently a piece of avant-garde art; the car is a classic. But the power of the images far outweighs the copy, particularly given the size of the text, which is very difficult to read. Owning a Lexus, one could gather from the ad, is analogous not only to owning the female body but to binding and gagging it as well.

These ads certainly illustrate the slippage between the female body and the auto body, promising, simultaneously, female power and masculine control of women, in a way that is remarkably similar to the Baker Electric ads of the 1910s—except now that control seems to be more technological than symbolic. Haraway suggests that "high-tech culture" challenges dualisms because "it is not clear who makes and who is made in the relation between human and machine" (177). Its potential for women lies in the fact that "up till now . . . female embodiment seemed to mean skill in mothering and its metaphoric extensions" (180). Now women can be identified with machines rather than dominated or threatened by them. It's a persuasive argument, yet looking at these ads should increase skepticism about female power in a cyborg world. Some of the machines pictured clearly evoke a high-tech feminine energy, challenging traditional female roles. But the mechanization of the female body also seems to facilitate its full or partial erasure, male ownership of what's left of it, and

Fig. 9. Lexus advertisement: "Are we the cutting edge of avant-garde?" (*Esquire,* 2000).

significant potential for the loss of female agency. These concerns become more pronounced when one considers the "appeal to mother" ads. The cultural baggage surrounding motherhood means that the maternal body is even more scripted than the
female body. Jennifer Scanlon, remarking that children often recognize name brands
before they have learned to read, concludes, "One could argue . . . that the language
of advertising or consumer culture is one of the first languages we master" (196). If
advertising is one of our first languages, then it is powerfully linked to maternal discourse. Mothers and ads function as a striking combination.

Volkswagen has long been known for its innovative ad campaigns and willingness
to poke fun at itself; the triumph of the VW Beetle in the 1960s partly stemmed from
a lighthearted self-deprecation that countered the seriousness of other automakers. By
the late 1970s we see further innovation in collapsing the distinction between woman
and mother. Pushing its Dasher station wagon in 1977, VW produced an ad with a
young woman announcing, "I bought a wagon out of wedlock." Playing to the stereotypes of the station wagon as a "mother's car," the ad tries to convince young women
that wagons are not just for mothers anymore. In thus challenging the assumptions
of what constitutes a mother's car, VW opens up the possibility that cars do not determine one's maternal status, that women buy cars regardless of their reproductive

capabilities. But the text goes on to rein in some of this potential with the line: "Some day I may load it up with kids. For now, I love to load it up and take off" (qtd. in Stern 123). It becomes clear that this young woman is simply a mother-in-training, a prudent woman buying her car of the future now.

If VW initially characterized itself as the lighthearted auto company, Volvo established itself as the choice for those who care about safety. In the February 1991 issue of *Harper's* magazine, Volvo takes a very hard line about the car a mother should buy (Figure 10). The fuzzy sonogram with its copy, "IS SOMETHING INSIDE TELLING YOU TO BUY A VOLVO?" eliminates the maternal body altogether. The figure of the fetus dominates the ad, and even the car is represented only by a very small image of a Volvo station wagon. Certainly, the ad plays on concerns about safety and protection. It is a mother's responsibility, it claims, to "listen" to her fetus while out car shopping, thus privileging the voice of the unborn over the voice of the mother. This car is not for her; it is for her fetus. As Janelle Taylor points out: "The suggestion is that the fetus will be safe only in the car. . . . What this means, of course, is that the 'maternal environment' herself is presented as potentially dangerous to the fetus, and its safety can be assured only by enclosing it (and her) within a car. The Volvo ad implicitly suggests that the pregnant woman is herself a sort of transport vehicle, and a relatively unsafe one at that: the steel body of a Volvo is needed to encase her, if the fetus is to remain safe" (78). The woman is only the first protective covering; the car offers real protection. This equation of woman and car is in many ways more troubling than the cyborg pictured in the Ford Focus ad. At least that seemed to be an intertwining of woman and machine; here the woman loses her maternity to a car.

The station wagon served as the mother's car of the 1960s and 1970s, but the 1980s and 1990s belonged to the minivan. Indeed, the development of the Dodge Caravan and Plymouth Voyager in 1983 helped to bring Chrysler Corporation back from bankruptcy. Marketed aggressively as a family car, the minivan replaced the station wagon, erasing VW's tentative suggestion that there is no such thing as a "mommy car." Advertising its Windstar minivan in 1999, Ford announced, "At Ford, we always listen to our mothers" (Figure 11). This ad goes on to brag that thirty members of the car's development team are mothers, without indicating the size of the complete team, making it impossible to determine the percentage of the maternal component. It then passes quickly over their knowledge of "torque converters" and moves to the more familiar body of maternal knowledge: diapers, infant seats, juice boxes, and security blankets. Ultimately, the ad tells us that mothers design cars to be nurseries on wheels, emphasizing safety above all else. The car is, after all, a minivan, a mother's car, a fact of which women are well aware. Denise Roy cites a woman in an online chat room complaining that driving a minivan "is like going around with a gigantic diaper bag

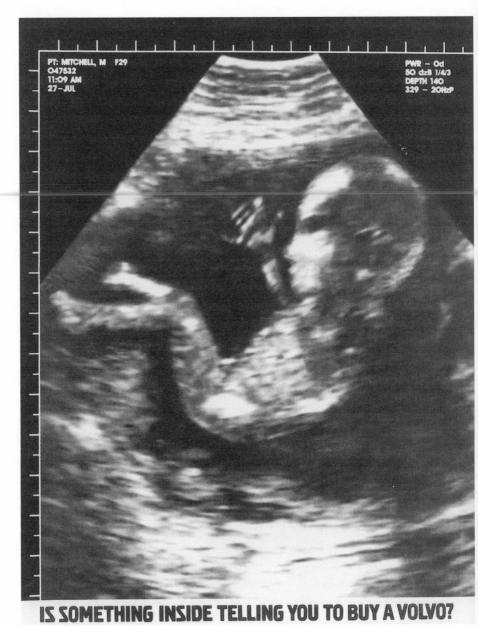

Fig. 10. Volvo advertisement: "Is something inside telling you to buy a Volvo?" (*Harper's Magazine*, 1991).

tied to my ankle" (10). Again, while there's ostensibly nothing to criticize in this concern over child safety, the construction of woman as mother and mother as defined through automotive technology remains troubling. An Oldsmobile ad appearing in *Good Housekeeping* in 2000 follows the same pattern, conceptualizing the car as replacing family and/or children's spaces: the family room, the remodeled basement, the

Fig. 11. Ford advertisement: "At Ford, we always listen to our mothers." (*National Geographic*, 1999).

Fig. 12. Oldsmobile advertisement: "Tree forts. Family rooms. Even remodeled basements can't compete." (*Good Housekeeping,* 2000).

tree fort (Figure 12). We get a rather distant view of the family sitting in the car, apparently watching TV, but Mom is prominently featured, providing food and drink. When automotive space becomes family space, it just means that Mom now has farther to walk to perform her domestic duties. It's not that the mother has become the car but that the mother sustains and services the car; she enables its current manifestation as family room.

A Saab ad in *Black Enterprise* in 2000 offers a somewhat more nuanced view, harking back to the VW ad from the 1970s (Figure 13). In setting up a distinction between Saab and parenthood—though the only featured character is a woman—it suggests that there may be a conflict between maternity and automobility, that automotive pressures on parents take all the fun out of cars. A new parent—a mother—"requires" a car that caters to her children's needs. The Saab, however, by adding a turbo engine,

Fig. 13. Saab advertisement: "Saab vs. Parenthood" (*Black Enterprise,* 2000).

"accommodates" both parents and children. Saab undoes the opposition between mother and car, but it does so by appealing to a standard male concern: engine size. Thus, Ford's reference to torque converters and Saab's appeal to engine size foreground women's exclusion from automotive technology—except for the technology that reinforces maternal identity. The "leaky distinction" between women and cars, then, may destabilize maternal identity, but it also extends it. Cars find new ways of enforcing maternity.

If advertising, as Twitchell suggests, "is more a mirror than a lamp" (111), then these ads mirror a culture in which women are either objects or mothers—and in some ways both. But the mirror of advertising has obvious flaws. "Advertising's mirror not only distorted," says Marchand; "it also selected" (xvii). These selections evoke a curiously nostalgic note, with the Oldsmobile ad looking back to the perfect nuclear family of the 1950s sitting around the TV set; the only difference being that now the TV is in the car. Toyota seems to want to take the women's movement back to the 1970s and sing protest songs, a relatively innocuous activity. Eliding the boundaries between the woman's body and the auto body is not intended to further Haraway's feminist agenda. Rather, the elision seems to create the myth not only of the continued dominance of the standard nuclear family but that cars help to reinforce this condition. As Don DeLillo suggests in his 1985 novel, *White Noise,* cars exemplify homogeneous family values. The narrator, Jack Gladney, watching the annual arrival of students to the "College-on-the-Hill," muses, "This assembly of station wagons, as much as anything they might do in the course of the year, more than formal liturgies or laws, tells the parents they are a collection of the like-minded and the spiritually akin, a nation" (3–4). The assembly of station wagons has accomplished what its advertisers hoped: to create a nation reliant upon the automobile for its definition. Looking at these ads makes clear, however, that something subtler than the creation of the "like-minded and the spiritually akin" is going on in automobile culture. While stereotypical families are certainly constructed by automobility, so is the maternal body, the ultimate source of that happy family. Automotive maternity simply makes a mother more the mother, revising Ford's 1912 claim that the car makes a woman more the woman.

Cars and Maternity: O'Connor, Didion, and Morrison

But many women novelists challenge these myths, presenting a very different picture of mothers and cars. As in the advertisements, maternal responsibility looms large, but automotive maternity comprises far more than concern over image and safety. Even without a slippage between mother and car, the mother's role nevertheless resonates as an integral part of automotive culture, though in far different ways from the lighthearted ads featuring women transporting popcorn to the vehicle or designing automotive nurseries. Flannery O'Connor, Joan Didion, and Toni Morrison explore the darker side of mothers and cars, the fears and guilt associated with the automobile's role in failed maternity.

The auto industry's attempt to persuade consumers that by buying minivans they could reconstruct the aura of the 1950s as an idyllic period of family harmony, when mothers stayed home except to ferry kids to various activities, falls flat in the face of the work of Flannery O'Connor. Writing during the late 1940s and 1950s, O'Connor is well aware of the pressure to conform to cultural norms: about family, about religion, and about cars. Her fiction makes clear the truth of the title of Stephanie Coontz's study of the 1950s, *The Way We Never Were*. If Americans never really did fulfill the 1950s fantasy of the perfect nuclear family with mother at home, dedicating her life to raising her children, then why has it lived on so stubbornly? The answers are many and complex, but one can clearly see the influence of the auto industry simultaneously to reinforce such myths and profit from them. One way to keep women restricted to maternal occupations is to convince them that business and technology are beyond them, to construct their automotive desires as focused on style and safety. Memos circulating within the Ford Motor Company in 1951 regarding information for an article in *Today's Woman* on "Women's Influence on the Automobile Market" indicate that, according to the company's surveys, women are generally "uninterested in the mechanical features of an automobile," with "prestige" being the "strongest interest" to women (Ralston 1, 3). Thus, in a news release in 1952 Ford declared that the company was "contradicting the myth that it's a man's world" by providing what women want: larger glove compartments, softer upholstery, and increased safety features such as "quicker-acting" brakes for women, who "are basically slower to make up their minds than men" (Ford News Bureau).

A 1956 article in the *Detroit News* noted that within the past thirteen years, with GM leading the way, the Big Three auto companies had all added women to their staffs "to watch out for the feminine interests." Those interests, of course, were perceived to be safety and style: "a woman wants a car that matches her home or her favorite dress," claimed Ford's token woman, Sally Somers (qtd. in Weiss 1F). In 1961 an Automobile Manufacturers' Association news release announced that the "little woman" demands door buttons that don't endanger fingernails, power assists for braking, steering, and adjustments, and family safety features. In attempting to balance style and safety, the auto industry implicitly acknowledged the split between appealing to a woman as woman and as mother. Can the woman concerned about her fingernails be entrusted to drive her children safely? Only, it seems, with help from the automobile industry. This acknowledgment highlights anxieties regarding the mother's role in the allegedly idyllic 1950s, a position that negotiates uneasily between wife and mother and, even more uneasily, between "nurturing motherhood" and "castrating 'momism'" (Coontz 32)—or, to cast it in contemporary terms, between the soccer mom and Susan Smith.

O'Connor's work capitalizes on tensions that the Big Three only hint at.[4] For her the automobile is far more than a vehicle for transportation, and its connection to women proves particularly problematic. Her most famous story, "A Good Man Is Hard to Find," details a car trip gone tragically wrong when the grandmother both misremembers and misrepresents the location of an old plantation house to manipulate the family into taking a detour that leads them into the path of the Misfit, a fugitive who appears, appropriately, driving "a big black battered hearse-like automobile" (20). A family in a car guided by the driver's mother is about to be annihilated by a psychopath in a hearse; clearly, it's no contest, physically or symbolically. No car safety feature can protect a family under the influence of the grandmother. In *Wise Blood*, O'Connor's first novel, Hazel Motes follows the path of his preacher grandfather, who "traveled three counties in a Ford automobile" (*Wise* 21). Hazel preaches from the hood of his Essex, his purchase of the car coinciding with his need to establish the "Church Without Christ." He seems to buy into the auto industry's claims that a car gives him place and standing in the world. That place, however, becomes painfully clear when he explains why this particular car is so important to him: "'It ain't been built by a bunch of foreigners or niggers or one-arm men,' Haze said. 'It was built by people with their eyes open that knew where they were at'" (126–27). The car represents for him white, able-bodied American masculinity. It provides him with an alternative home, a space away from the dominance of women.

When a policeman wrecks the dilapidated vehicle by pushing it off an embankment, Motes blinds himself, relegating him to the world of the disabled, not of virile car builders. He consequently gives up his proselytizing, his faith somehow connected to his car. But loss of the car does not only entail the loss of the Church Without Christ; it sends Motes back to the world of women: Sabbath Hawks and his landlady, Mrs. Flood, both of whom hope to marry him. Yet while their ostensible interest in Motes is as a husband, both of them, particularly Mrs. Flood, see themselves as his caretakers. "It's fortunate for you," says Mrs. Flood, "that you have this warm place to be and someone to take care of you" (226). With this emphasis on care it is easy to conclude that losing the car delivers Hazel Motes back to the mother. Ultimately, submersion into the female world proves to be even more terrifying than life without a car, and Hazel lies down in a ditch to die. Cars thus seem to provide some separation from the mother; when that is gone, so are life, manhood, and faith—at least for those preaching the word of the Church Without Christ.

Obviously, cars take on a wide range of meaning in O'Connor. Roger Casey argues that in her work "the automobile is a place to be, a hierophanic object, a vehicle for escape, an embodiment of the new, an object of materialism, an emblem of the duality of the body and the spirit, a symbol of the depersonalization of humanity through mechanization, a locus for transformation, a structural device, a metaphor for characterization, a microcosm for society, a symbol of isolation and insulation, a deliverer of prophecy, and a place for sex" (107). This constitutes quite a catalog. And in her 1953 story "The Life You Save May Be Your Own" the car functions as an emblem of failed maternity. While O'Connor does not elide the categories of woman and car, she explodes any mythology of the mother as cheerful family chauffeur and of women's automotive interest being restricted to style and safety.

The mother in the story, referred to as "the old woman," justly challenging her status as mother, basically bribes a traveling carpenter to marry her deaf—and probably mentally disabled—daughter by offering him a car. Mr. Shiftlet's interest in the car is clear from the start, as O'Connor deftly inserts references to it as he makes conversation with the old woman. His eye is on the car, and the old woman knows it. He accepts her offer to sleep in the car and be fed in exchange for work; in fact, he's delighted by it, pointing out that "the monks of old slept in their coffins!" ("Life" 59). By comparing himself to monks and their habits of reminding themselves of their own mortality, Mr. Shiftlet situates the car as a spiritual space, a reminder of his eventual end. As he later asserts, "The body, lady, is like a house: it don't go anywhere; but the spirit, lady, is like a automobile: always on the move, always" (63). He restores the car, driving it out of the shed "as if he had just raised the dead" (61). The wording reveals Mr. Shiftlet's hubris in likening automobile culture to Christianity and himself to Jesus. Hazel Motes may preach for the Church Without Christ, but Mr. Shiftlet sees himself as Christ. He seems to think he has saved his soul by reincarnating the car, but O'Connor thinks very differently, evidenced by the way she situates him in relation to the car.

Mr. Shiftlet comes off as comically ignorant regarding car culture. He may be able to fix the car, but he, like Hazel Motes, knows nothing about what goes into making a car, despite his claims to the contrary. Having decided that the car must be a 1928 or 1929 Ford, he "said he could tell that the car had been built in the days when cars were really built. You take now, he said, one man puts in one bolt and another man puts in another bolt and another man puts in another bolt so that it's a man for a bolt. That's why you have to pay so much for a car: you're paying all those men. Now if you didn't have to pay but one man, you could get you a cheaper car" ("Life" 60). Any Ford built in 1928 or 1929 would have come off the assembly line, cheaper by far than relying on one overall builder. And the vast majority of the U.S. population, more attuned to automobile history than Mr. Shiftlet, thus sees him for the fraud he is. He worships something he doesn't understand. By subtly mocking Mr. Shiftlet's grasp of the auto industry, O'Connor positions him in a place often occupied by women: outside of car culture. She also makes clear that his religious failure is linked to the car. One cannot ignore the workings of the material world and expect to find grace. In contrast, the old woman may know nothing about cars, but she is a shrewd mother who knows exactly which buttons to push in order to unload her daughter. She plays on his obvious desire for the car, agreeing to buy a fan belt for it and offering him money to get it painted, if he'll marry her daughter. The car thus provides the mother with an invaluable tool in dealing with her situation. Rather than being shaped by car culture into a stereotypical 1950s mother, she exploits the car to get what she wants: an end to maternity. She perceives the car as far more than upholstery and door buttons.

In fact, she capitalizes on a slippage not between mother and car but daughter and car; she offers Mr. Shiftlet her daughter, who comes with a car attached. Mr. Shiftlet "had always wanted an automobile but he had never been able to afford one" (65), and so he somewhat reluctantly agrees and drives off with daughter, Lucynell, whom he then abandons at a roadside café. He cannot, however, leave the mother behind so easily. Possibly guilty over deserting Lucynell, whom the counterman describes as looking "like an angel of Gawd" (66), Mr. Shiftlet picks up a hitchhiker and announces, out of nowhere, "I got the best old mother in the world so I reckon you only got the second best." His mother, he goes on to insist, "was an angel of Gawd" (67). In thus conflating his mother, whom he leaves, with Lucynell, whom he leaves, Mr. Shiftlet tries to justify his actions by covering them over with a load of sentimentality. His dedication to his mother is supposed to erase his callousness toward Lucynell in exploiting her to get the car; because both are "angels of Gawd," he can subsume Lucynell in a stereotyped panegyric on his own mother and claim a connection to "Gawd" through his mother. But this is O'Connor, and of course it doesn't work. What he wants, we realize, is not a mother but her car. The boy he picks up turns on him, screaming, "My old woman is a flea bag and yours is a stinking polecat!" (67). By refusing to allow Mr. Shiftlet to find absolution in sentimental maternal pieties, O'Connor keeps intact the more disturbing picture of the mother as automotive manipulator.

One cannot idealize maternity in this story, in which mothers use cars to sacrifice their daughters. While the old woman demonstrates some love and concern for her daughter when the couple drives away, she cannot be ignorant of the probable outcome. Even in the 1950s, allegedly the highpoint of the reign of the perfect mother, O'Connor makes clear that cars do not make a mother more the mother. Nancy Clasby has identified the car in this story as a "mechanical womb" (514), tying it strongly to maternity yet also implying the continuation of the perception of mother as carrier.

I would argue, however, that the car enables this woman essentially to undo her maternity, calling into question both the mother's humanity—and thus opening the way for a kind of inhuman cyborg maternity—and the cultural imperative that says mothers use cars to enhance their motherhood, that mothers are only transporters. Here the car transports one away from the mother. Mr. Shiftlet saves the car, at the probable cost of his own salvation, leaving mother and daughter behind. He, like Hazel Motes, sees the car as an escape from women, but O'Connor ties it strongly to maternity. She thus capitalizes on the tension between mother as nurturing and mother as smothering; she offers salvation and leads him to damnation. He may drive away from the mother, but he's driving the mother's car. O'Connor, so often read through the lens of Catholicism, also demonstrates considerable understanding of the world of technology and its place in modern spiritual life.

It seems to be a big jump from O'Connor's challenge to the perfect mother of the 1950s to Joan Didion's 1970 novel, *Play It as It Lays,* set at the apex of car culture: Los Angeles in the late 1960s. Yet Didion offers a very different spin from dominant images of scantily clad women draped over automobile hoods. She, like O'Connor, focuses on the mother's position in automotive culture. While the novel certainly contains its fill of sleek, sexy cars, Maria Wyeth, a minor actress, drives the California freeways as a desperate means of staying sane and so gaining more say over the fate of her daughter, who has been shut up in a mental institution. Moving back and forth between Maria's own residence in a psychiatric home and the events leading to her commitment, the narrative situates the car as Maria's anchor in the arid Hollywood landscape. Maria's strongest tie is to her daughter; she cooperates with her therapists for Kate's sake: "I bother for Kate. What I play for here is Kate. Carter [her husband] put Kate in there and I am going to get her out" (4). Thus, amid the glamour of Hollywood, the sex, the drugs, and the alcohol, Maria defines herself primarily as a mother. In trying to establish an identity outside of the dominant Hollywood culture, Maria uses the car to forge a woman's—and a mother's—alternative to the sexist and masculinist California car culture.

Maria drives to cope with the breakup of her marriage and the removal of her daughter. It "was essential (to pause was to throw herself into unspeakable peril) that she be on the freeway by ten o'clock. Not somewhere on Hollywood Boulevard, not on her way to the freeway, but actually on the freeway. If she was not she lost the day's rhythm, its precariously imposed momentum" (15). Driving gives shape, meaning, and safety to her days, staving off "unspeakable peril." We do not see her become one with the car so much as we see her look to the car for a sense of who she is. Rather than incorporating her into family life, the automobile serves almost as a retreat where she can recoup and regain her position as mother. She needs to reassure herself of her competence as a driver in order to perceive herself as a mother; successfully negotiating the roads serves as a necessary prerequisite to regaining custody of her daughter. "Again and again she returned to an intricate stretch just south of the interchange where successful passage from the Hollywood onto the Harbor required a diagonal move across four lanes of traffic. On the afternoon she finally did it without once braking or once losing the beat on the radio she was exhilarated, and that night slept dreamlessly" (16). To be a success on the road suggests the potential for success as a human being, a precondition for success as a mother.

The car's role in reinforcing her sense of self is so powerful that she doesn't even need actually to be in it. A hypnotist trying to help her encourages her to imagine a womblike condition: "You're lying in water and it's warm and you hear your mother's voice." But Maria refuses his vision, insisting that in her own mind, "I'm driving Sunset and I'm staying in the left lane because I can see the New Havana Ballroom and I'm going to turn left at the New Havana Ballroom. That's what I'm doing" (124). By replacing the womb with the car, Maria implicitly acknowledges that cars rather than mothers hold sway in our culture, anticipating the message of the 1991 Volvo ad. When she looks for a place of comfort, she seeks the car. Thus, while the mother may not have become the car, she seems to have been replaced by it. Indeed, Maria's obsession with the automobile becomes uncomfortable, reminding us of the effects of displacing human identity with automotive identity. The car may help her to cope with the loss of her daughter, but, by usurping the role of the mother, it also undermines her attempts to reclaim her maternity. If the car serves as the mother, why do we need "real" mothers, particularly those unable to retain custody of their children?

The displacement of the mother by the automobile is even more pronounced when Maria unwillingly seeks an abortion. Coerced by her estranged husband, who threatens to take Kate if Maria does not comply with his demand that she terminate her pregnancy, she goes through an elaborate procedure in this pre–*Roe v. Wade* era to arrive at the unknown doctor's secret location. She must first meet a man in a parking lot, who will direct her further. Their conversation on the drive is of cars. Asking about her gas mileage, the man ponders trading in his Cadillac for a Camaro, describing the one he has in mind and talking about how to get it for the lowest price. The focus on cars helps her to go forward with the procedure: "In the past few minutes he had significantly altered her perception of reality: she saw now that she was not a woman on her way to have an abortion. She was a woman parking a Corvette outside a tract house while a man in white pants talked about buying a Camaro" (79). Maria may be comforted, at one level, by displacing the abortion with car talk. But an abortion is not something that can be erased, and she clearly grieves throughout the remainder of the novel, the loss adding to her guilt and desperation. The material reality of the female body prevented from carrying a pregnancy to term, a woman denied the opportunity to become a mother, reminds us that though cars may displace the mother, they cannot ultimately become her. No car is faced with the agonizing choice between a living child and a fetus, and the raw emotion involved in such a decision is anything but mechanical. As much as Maria tries to recast herself from "woman seeking abortion" to "woman discussing cars," she fails, sitting in her car and crying on "the day the baby would have been born" (141).

Didion thus offers a very ambivalent picture of the link between mother and car. Maria's experiences open up an alternative automotive world, one in which women seek to control their lives by driving, to replace the womb with the car, and to cover over abortions with car talk. But this hardly comes across as liberatory, particularly given that within the frame of the novel, at the time of telling, Maria is committed to a psychiatric hospital and is not driving anywhere. What we get here is an incomplete vision of what a maternal cyborg might look like. Didion's novel, situated in the late 1960s, comes at a time when women were expected to use cars to extend their maternal roles, to ferry kids to various activities and shop for the family. Didion, like O'Connor, challenges the apparent beneficence of this image, starting to pry open the Pandora's box containing mothers and cars, opening a road for later writers to drive through.

Toni Morrison takes up the challenge. At the start of her 1999 novel *Paradise* a group of men from the all-black town of Ruby head out to murder a group of women living in a nearby house called the Convent. The men, seeing growing evidence in their town of the intrusion of contemporary American culture, largely in various forms of teenage rebellion, decide that the only explanation for the infiltration of new ideas must be the women in the Convent, women who "call into question the value of almost every woman [they] knew" (8). Ironically, these men see themselves as protectors of women; the Morgan twins recall an incident when outsiders drove through and terrorized a group of young girls by circling them with their cars and exposing themselves through the windows. The men of the town gathered about with rifles, their presence enough to scare off the intruders. This memory does not identify the girls in question; it is very specific, however, when it comes to the details of the cars. "Three cars, say, a '53 Bel Air, green with cream-colored interior, license number O85 B, six cylinders, double molding on rear fender pontoon, Powerglide two-speed automatic transmission; and say a '49 Dodge Wayfarer, black, cracked rear window, fender skirts, fluid drive, checkerboard grille; and a '53 Oldsmobile with Arkansas plates" (12). Morrison replicates the language of automobile ads here, reflecting the extent to which it permeates American culture. When we talk about cars, we become representatives of the auto industry. The men driving these vehicles remain nameless and faceless, almost as if the perpetrators are the cars themselves. The language, in fact, also reads a bit like a police blotter, detailing the description of the cars to alert others to be on the watch for them. The cars are important; the men, seemingly, are not. Or maybe, given that such behavior seems incomprehensible, one focuses on the knowable details of the scene: the cars. By concentrating on automotive details, it may be possible to defeat the potential sexual predators who drive them. Regardless of how one interprets these details, what stands out are the cars. They are defined, described, and, finally, defeated.

Cars, then, do not constitute the external threat to Ruby, but they certainly serve as the vehicles through which such threats enter the community. Thus, it is telling that the next section of the novel tells the story of a woman and a car, both of which end up at the Convent. Mavis, whose infant twins have smothered in her partner's Cadillac while she runs into the store to buy hot dogs, finds comfort in the car in which they died. She wants to sleep "in the back seat, snuggled in the place where the twins had been, the only ones who enjoyed her company and weren't a trial" (25). The car, site of her failed maternity, also allows her to maintain maternal identity. While it does not resurrect the twins, it does seem to reincarnate her motherhood. She steals the car as a means of prolonging that role, even though in committing such an act, she leaves behind her living children, convinced that they now wish to kill her. Unlike Didion's Maria, who sacrifices the unborn for the born, Mavis privileges the unliving over the living, calling into question what constitutes maternity. Morrison's presentation of Mavis's confused sense of motherhood reminds us of the complexity that the automobile ads erase. What happens when irresponsible car behavior leads to the death of one's children? Is one still a mother? What does it mean to be a mother of dead children? This is a question that Morrison comes to again and again in her fiction, most famously in *Beloved*. But it is only in *Paradise* that she links it to the automobile.

By connecting the mother to the car, Morrison almost redeems the pain of killing a child so powerfully portrayed in *Beloved*. Because Mavis seems to feel the presence of her twins in the Cadillac, she is able to maintain her feeling of herself as mother. Rather than presenting a happy family doing all in their power to guard their children by buying expensive cars and elaborate child protection systems, as portrayed in contemporary ads, Morrison employs the automobile in a far more complex manner. It allows for an alternative form of maternity, one that continues after death. Thus, while the mother does not necessarily become the car, the whole notion of maternity is rather eerily portrayed as somehow enabled by the very car that kills the children. In eliding the boundaries not so much between woman and car but between life and death, Morrison draws on some of the imagery of cyborgs as both human and inhuman but grounds that concept very firmly in the material reality of a slightly crazed woman trying desperately to erase her babies' tragic deaths. Mavis clings to the car in an effort to perpetuate what the vehicle has already taken away. The car's power over life and death transcends that of the mother; the best she can do is to sneak away in the all-powerful Cadillac, her position now reduced to the care and feeding of the car. Her efforts to find the money to buy gas and to pay for a paint job to disguise it are presented in considerable detail. Mavis now mothers the car, constructing a very different vision of maternity and family.

This car, in fact, has been presented as a violation of the family and community: "The neighbors seemed pleased when the babies smothered. Probably because the mint green Cadillac in which they died had annoyed them for some time" (21). The intrusion of a luxury car—even a used one—disrupts the neighborhood, cautioning against the privileging of cars over community. But the neighbors' resistance goes beyond simple jealousy.

> When Frank bought it and drove it home the men on the street slapped the hood and grinned, leaned in to sniff the interior, hit the horn and laughed. Laughed and laughed some more because its owner had to borrow a lawnmower every couple of weeks; because its owner had no screens in his windows and no working television; because two of his six porch posts had been painted white three months earlier, the rest still flaking yellow; because its owner sometimes slept behind of the wheel of the car he'd traded in—all night—in front of his own house. And the women, who saw Mavis driving the children to the White Castle . . . flat-out stared before shaking their heads. As though they knew from the start that the Cadillac would someday be notorious. (28)

Morrison here may be playing upon the cultural stereotype of the welfare Cadillac, the expensive car purchased by the financially disadvantaged. While she appears to evince some disapproval of the purchase, in typical Morrison fashion she explores the lure and meaning of the car. She focuses not on financial judgment but on how the car, once purchased, positions its owner familially and culturally.

The Cadillac is a far cry from providing an alternative tree fort or family room. It divides rather than unites because of the reminder of everything that this family does not have: stability. Yet just as women in the early part of the century prized the car over the bathtub, Frank values the car more than window screens and television. Even Mavis "loved it maybe more than he did" (25). This is no stereotypical scene of the husband-father sacrificing his family's needs for a fancy car, for Mavis accepts and participates in the situation and, in fact, chooses the car above her remaining children, a greater offense against family than doing without TV. This car functions as far more than a status symbol; it both erases and reincarnates maternity. While the car enables a kind of posthumous maternity, it also undoes "real" motherhood as Mavis adopts automotive maternity rather than real-life motherhood.

The car in the novel thus seems to weaken one's position in the family and in the community. Rather than encouraging participation, it creates isolation. Yet Mavis, once she has stolen the car, begins to undo the isolation the Cadillac has imposed. She picks up women hitchhikers both for help with gas money and also for company. The "quiet" after the last one leaves is "unbearable," and Mavis begins to hallucinate visions of Frank hunting her down (35). Being alone in the car becomes impossible, reminding her of her failures rather than perpetuating her maternity. In order to restore the Cadillac's connection to maternity, she must use it to re-create a sense of family by situating it within a community. Once she arrives at the Convent and the car helps to support the household that was previously without any transportation, Mavis can begin to construct an alternative family in the Convent, using the car to provide greater convenience for the women. The car that helped to destroy the nuclear family creates the possibility of forging alternative families. It also undoes the segregation between the Convent and Ruby. The Convent women drive to K.D. and Arnette's wedding; they find scant welcome when they arrive, possibly due to the realization that women with wheels cannot be counted on to remain seventeen miles distant. The car, then, muddies the separation between patriarchal Ruby and the matriarchal Convent. As Magali Michael notes, "The novel presents the community of women . . . as able to negotiate a separatism that is merely temporary, that is constructive and inclusive, that reconceptualizes power as power *with* rather than power *over*, and that depends on dynamic coalition practices" (650). None of these negotiations or coalition practices would be possible without the presence of the Cadillac. One needs the material reality of the car to bring the world of maternity into culture, into Ruby.

But cars function differently for men. Deacon Morgan, who drives his car three-quarters of a mile to work every day, justifying such "silliness" by the "weight of the gesture," also refuses to allow anyone else access to it: "His car was big and whatever he did in it was horsepower and worthy of comment." The community remarks on how he "washed and waxed it himself—never letting K.D. or any enterprising youngster touch it; how he chewed but did not light cigars in it; how he never leaned on it, but if you had a conversation with him, standing near it, he combed the hood with his fingernails, scraping flecks he alone could see, and buffing invisible stains with his pocket handkerchief" (107). Rather than integrating him into the community, the car announces "the magical way he (and his twin) accumulated money." Mavis, who sees the car as an extension of maternity, gathers other people around and in it; Deek, who views it as evidence of his "prophetic wisdom" and success, his superiority, uses it to impress people, to separate himself from the "contamination" of the women of the Convent.

This is not the first time a car has functioned as emblem of masculine success in Morrison's work; she explores the role of the car within the community much earlier in her career as well. In *Song of Solomon* she presents the isolation of Macon Dead by describing his automotive behavior: "Macon's wide green Packard belied what they [the African-American community] thought a car was for. He never went over twenty miles an hour, never gunned his engine, never stayed in first gear for a block or two to give pedestrians a thrill. He never had a blown tire, never ran out of gas and needed twelve grinning raggle-tailed boys to help him push it up a hill or over to a curb. . . . No beer bottles or ice cream cones poked from the open windows. Nor did a baby boy stand up to pee out of them. . . . [T]he Packard had no real lived life at all. So they called it Macon Dead's hearse" (32–33). To some extent the community envisions the car in much the same light as that offered by the auto industry: a vehicle that incorporates a person into the community, albeit in very different ways. Macon's refusal to participate in the excitement of automobility, his insistence on seeing it as an item of conspicuous consumption, and, ultimately, his inability to see it as a living entity all mark him as a man without a "real" living identity and family: in short, he's Dead. Although her use of names is somewhat less overt in *Paradise*, the message is similar: Deacon Morgan, whose significant difference from his twin brother—at this point in the novel—lies in their disputes over Chevrolets versus Oldsmobiles, is shaped by the "weight" of his car, a burden that eventually drags him down to murder (155). Mavis's car enables a new form of family; Deacon's car destroys it.

Thus, while cars may destabilize maternity, unleashing its connection to life and opening it up to the possibility of a form of cyborg maternity, both human and inhuman, they also remind us of the fragility of family and community ties. So identified is the car with the Convent women that when Roger Best arrives to collect the dead after the assault, he finds neither bodies nor Cadillac. The fate of women in the final section—whether materially present or ghostly—is shared by the car. By placing the car and the women between the boundaries of the real and the unreal, Morrison acknowledges the cyborg power of the automobile and its integral role in reformulating community. It seems to replace the nuclear family, allowing Mavis to continue her dead family and Deacon to become a man who challenges family, refusing to allow his nephew to marry a girl he has impregnated, refusing to sully his sense of pride in his own family identity. The men of Ruby may welcome automobiles—as opposed to television—as part of their community, but by accepting the car, they must also confront its larger implications: the inhuman maternity that they see as such a threat in the Convent. Despite their desires to keep the town firmly linked to its past, the very mothers they seek to protect have learned to take advantage of the age of the automobile, driving to distant hospitals to take advantage of increasing medical technology to give birth and, as the midwife Lone puts it, denying her their wombs (271). The mothers of Ruby rank automobility and the access it offers to a wider world above the tradition of giving birth at home. This link between mothers and cars highlights the similarity between the women of Ruby and the women of the Convent, reminding the reader that the men's opening attack on the outsiders might be better directed against their own wives, if they hope to defeat the encroaching female power from driving into their town.

Erdrich and Maternal Cyborgs

The automotive age flourishes in Louise Erdrich's fiction, possibly more than in that of any other writer considered in this study. Cars are everywhere in her work, functioning, as Roger Casey contends, as "special, even sacred, possessions" (148). Her first novel, *Love Medicine,* is structured as a series of interlocking stories about a Native American community, one in which cars play significant roles. Functioning as a bridge between living and dead, the automobile enables a way of the life that allows for the continuity of Native American tradition within the technological present of American culture. In "The Red Convertible" Lyman Lamartine tells of the red Oldsmobile convertible he and his brother Henry acquire and drive north to Alaska. Their first sight of it is staggering: "There it was, parked, large as life. Really as *if* it was alive" (*Love* 182). Again, the distinction between human and machine is elided as the car looms "large as life." After Henry returns from Vietnam, unable to settle down and assimilate back into the family and community, Lyman decides that the one thing that might help to "bring the old Henry back" is the car (187). So, he takes a hammer and goes out to destroy it, hoping to galvanize Henry into action once he sees the damage. The plan seems to work; Henry discovers the car's condition and spends all his time fixing it up. As Lyman notes, "He was better than he had been before, but that's still not saying much" (188). And indeed, it turns out to be a temporary solution. Once Henry has the car back in shape, he realizes he cannot go on, walks into the river, and drowns. Lyman sends the car in after him as a final gesture of love and farewell. Here we see the car taking on cyborg characteristics, part human, part machine. Although we tend to think of Indians as focused on the natural world, it is the car that embodies life to Lyman and Henry, so once Henry's life is through, Lyman cannot bear to let the car live on.

This episode may strike some non–Native American readers as slightly odd; cars

Fig. 14. Geronimo in car. U.S. National Archives (75-IC-1).

and Indians seem to be, at some level, mutually exclusive. Native Americans are associated with horses in the U.S. cultural imagination, not cars. Yet the assumption that cars are somehow inimical to Native American life has been forcefully challenged by Philip Deloria in *Indians in Unexpected Places,* which has a fascinating chapter on Native Americans and cars. He begins with a 1904 photograph of Geronimo in a car (Figure 14), arguing that to "imagine Geronimo riding in a Cadillac . . . is to put two different symbolic systems [primitivism and modernity] in dialogue with one another" (136). Jimmie Durham, commenting on the same photo, points out that in this picture Geronimo was a prisoner of war, "not allowed a rifle or his chaparral." Instead, "he puts on your hat, takes the wheel, and stares the camera down" (58). In other words, Geronimo in a car is still Geronimo the warrior, if anything a little more threatening than if he could be dismissed as an image of the stereotypical Indian. But Indians, as Deloria claims, "have lived behind the wheel in all kinds of powerful ways" (139). The advent of cars corresponded with new influxes of cash on reservations—from claims cases and sales of allotments—and much of that money found its way to auto dealers. Viewed by some observers as a move toward progress and civilization and by others as "squandering" money on shiny trinkets, Native Americans themselves, says Deloria, saw things differently. "In truth, automobile purchase often fit smoothly into a different logic—long-lived Indian traditions built around the utilization of the most useful technologies that non-Indians had to offer" (152).

Joy Harjo confirms the place of the car in Native American culture. Referring to a photograph of her great-grandparents in a Hudson purchased with oil money, she muses: "This picture . . . explodes the myth of being Indian in this country for both non-Indian and Indian alike. I wonder how the image of a Muscogee family in a car only the wealthy could own would be interpreted by another Muscogee person, or by another tribal person, or by a non-Indian anywhere in this land. It challenges the popular culture's version of 'Indian'—an image that fits no tribe or person. By presenting it here, I mean to question those accepted images that have limited us to cardboard cut-out figures, without blood or tears or laughter" (91–92). It is only by considering the car as an integral component of Native American culture that one can begin to look beyond the "cardboard cut-out figures." Furthermore, given the geography of many reservations, with miles of open space, "automotive mobility helped Indian people evade supervision and take possession of the landscape, helping make reservations into distinctly tribal spaces" (Deloria 153). Thus, Lyman's use of the red convertible as a means first of trying to cure his brother and then as part of his burial goods reflects a long-standing tradition about the integral role the automobile plays in Native American communities.[5] Deloria cautions, "Rather than succumb to the powerful temptation to imagine Indian automobility as anomalous, we might do better to see it in Indian terms—as a cross-cultural dynamic that ignored, not only racial categories, but also those that would separate out modernity from the Indian primitive so closely linked to it" (155–56).

Deloria's primary focus is on modernity and Indian participation in technology. He dwells only briefly on gender, saying that Indian men, constructed as feminized beings, "were doubly or trebly unsuited to the automobile," while "Indian women, caught up the mingled terms of race and gender, might be imagined outside the automobile even more easily" (146). Erdrich's work in some ways reflects the position of Indian women as outside automobility; men are the primary drivers and car owners: Henry and Lyman with their red convertible, and Lipsha Morrissey in *The Bingo Palace* with his bingo van. But *Tales of Burning Love* provides a significant exception. The central context for the story is a group of four women trapped in a car during a blizzard after the funeral of their mutual husband, Jack Mauser, telling each other stories of their lives with Jack. In particular, Dot, the driver, has had other interesting driving experiences, having played chicken with a man who exploited her, using her compact car to force him and his truck into a ditch. Eleanor, another of the women, underscores the significance of the episode by dwelling on it: "I want to try to understand what you've just told me. At that moment you—a grown woman with a dependent child—decide to play chicken, in a compact car, with a Mack truck?" (*Tales* 93).

What I find striking here is the implication that Dot's motherhood, if nothing else, should prevent such recklessness. Granted, her daughter is not in the car with her, but her position as mother should nevertheless restrain her, should construct her automotive behavior. Mothers in cars must behave. There are no ads featuring mothers playing chicken. In fact, mothers are not even supposed to be driving compact cars; Dot's car, however, reminds us of the financial reality of the majority of mothers in this country: compact cars may be all they can afford. Thus, the episode also reminds us of the privilege inherent in car culture: a mother who drives a compact car is a bad mother, putting her children unnecessarily at risk. As an article in *Going Places,* the journal of the American Automobile Association, cautions: "Some vehicles, particularly compact and sports cars, may not work for you if you're pregnant. You may find that, like your wardrobe, your vehicle may have to increase in size as well" (Fischer 10). Compact cars are not for mothers-to-be or mothers. Rather, they should be driving Volvos, minivans, or, increasingly, SUVs. By 2002 SUVs had captured the largest share of women buyers, with 23.9 percent of the market versus 23.2 percent for mid-size cars. Roughly 35 to 40 percent of all SUV purchases are now made by women ("SUVs"). Ironically, this does not automatically enhance one's nurturing capacity. In fact, the SUV, like Susan Smith, reminds us of destructive automotive power. An article in an online journal, *Freepressed,* in December 2003 cites Hummer—giant-sized off-road vehicles developed from High Mobility Multipurpose Wheeled Vehicles (HMMWV) used by the military—and SUV owners responding to criticism that such vehicles endanger others by crushing smaller cars in accidents. The speaker, identified as a "soccer mom," says her Cadillac Escalade, a large luxury SUV, makes her feel "safe and secure." If other drivers don't feel that way, she says, "they can go buy their own SUV" ("Hummer Owners"). The Escalade retails for roughly fifty thousand dollars; if owning one is necessary for good mothering, most mothers are in real trouble. This is a long way from Mother Nature, particularly given the appalling environmental impact—and even the safety record—of the SUV.[6]

But Erdrich presents a very different picture of mothers and cars. Industry-constructed visions of automotive maternity find scant attention in her fiction, from Dot, who shows no remorse, to June, whose story begins in *Love Medicine* and continues through *Bingo Palace* and *Tales of Burning Love.* At the start of *Love Medicine* June walks out of a truck—Jack Mauser's—off into a snowstorm and freezes to death. Erdrich carefully focuses her imagery on June as body and June as soul: "Even when her heart clenched and her skin turned crackling cold it didn't matter, because the pure and naked part of her went on. The snow fell deeper that Easter than it had in forty years, but June walked over it like water and came home" (7). There is nothing automotive in this episode. Replete with Christian imagery, the passage illustrates the transformation of June from body to soul. The technological associations come later, when King, June's oldest legitimate son, cashes in the money her life insurance policy pays out and buys a new car—thereby fulfilling, rather morbidly, the promise of *My Mother the Car.* The family consequently associates her with the car; Eli, her foster father, won't ride in it because it "reminds him of his girl" (23). The woman may have become the car, as in the Corolla ad from *Better Homes and Gardens,* but this is hardly presented as a desirable transformation. No one other than King and his wife "seemed proud of it . . . . Nobody leaned against the shiny blue fender, rested elbows on the hood, or set paper plates there while they ate" (24–25).

The assumption is that cars should be integrated into family life, an extension of family identity. People should lean against them and eat off the hood. Cherokee writer Louis Owens confirms this point as he recalls:

> In photograph after photograph, my mother and father, aunt, uncle, or grandmother, my pudgy baby self, my brothers and sisters and cousins, are sitting on the chrome bumper or shiny fender or hood of a thirty- or forty-something Chevy or Ford or Buick. . . . Always there is family associated with each automobile, and an almost tangible sense of pride in that association. Adult hands casually touch the chrome and caress the shining hood and shoulders lean confidently into sturdy metal, feet possessively lifted onto the running boards. Only one house holds in my memories of those photographs . . . but I remember what seems like scores of sleek, sloping-fendered chrome-edged cars and an obvious desire to be associated with those vehicles.   (160–61)

What is striking about these scenes of family cars is how different they are from the ads; what we see is not mother schlepping popcorn and juice boxes out to this new family space but extended families creating an interactive environment around the car, similar to the assumption in Morrison's work that cars foster extended family communities. Mothers are not singled out as bearing the automotive burden.

But when the mother becomes the car, things change. Rather than celebrating family, in *Love Medicine* King's car transforms human into cyborg. The family clearly judges him as callous for spending his mother's life insurance money on a car, and the car becomes eerily human and thus something to be avoided. By collapsing mother and car, King sacrifices an extended family connection to the car. He wants exclusive control of the mother, refusing to allow others to drive the car, thereby betraying Native American family tradition. When his wife locks herself in the Firebird to escape his drunken rage, he "threw his whole body against the car, thudded on the hood with hollow booms, banged his way across the roof, ripped at antennae and side-view mirrors with his fists, kicked into the broken sockets of headlights. Finally he ripped a mirror off the driver's side and began to beat the car rhythmically, gasping" (*Love* 35). This degree of violence reveals, Casey suggests, "the anger he feels over his mother's senseless death. The car thus serves as a substitute on whose body he expresses his suppressed feeling of anger about and towards his mother" (150). The car may be a material substitute, but it nevertheless functions as a virtual mother: to assault the car is to assault the mother. The leakage between June and the car hardly suggests an empowering combination of human and machine. It has more in common with the Lexus ad from 2000 as a means of finally silencing June and bringing her under male control. When the mother becomes a car, it reverses the parent-child hierarchy, giving the son command. While it certainly fulfills Haraway's premise of an alternative female embodiment, it also erases the mother and weakens the Native American family.

Except that it really doesn't, for given King's out-of-control rage, it becomes clear that turning one's mother into a car hardly leaves the son in control. More important, this isn't the last we see of the car, as the novel circles back around to it at the close, when King loses it in a poker game to his unacknowledged half-brother, Lipsha Morrissey. Lipsha, who has never forgiven June for abandoning him as a baby, finds a release in the vehicle: "I thought I would never quit driving, it felt so good" (*Love* 361). Driving his newly acknowledged father, an escaped convict, to freedom, Lipsha experiences a kind of epiphany, facilitated by the car and experienced through the car. In fact, the car rebirths him—and his father. Opening the trunk in which Gerry has hidden, Lipsha sees him "curled up tight as a baby in its mother's stomach" (362). The car, we are forcefully reminded, is the mother. And in thus helping to birth his own father, Lipsha aligns himself with his mother. As he approaches home, he finally admits, "There was good in what she did for me, I know now" (366). The car reconciles him with his mother both because she *is* the car and he's now driving her and also because in driving the car he begins to understand and forgive her. He now finds reinstatement in the family, having connected to both mother and father. Then, in the novel's final lines, Lipsha notes: "The morning was clear. A good road led on. So there was nothing to do but cross the water and bring her home" (367). Erdrich's language here carefully echoes the earlier description of June's death, especially since both episodes claim to bring her home. But here, rather than walking on water to her death, June, as car, is driven over the water to her home. For Erdrich's characters, Casey says, "the car functions sacramentally, sacrificially, spiritually, and as sanctuary" (154). In this closing section the cyborg, June/car, effects a sense of healing. While it may not indicate any particular sense of female empowerment—after all, it is Lipsha, June's son, who finds peace—embodying June in the car rather than in Christ walking across the water nevertheless conveys a means for expressing identity in a world in which, even on Native American reservations, technology shapes human life. June as Christian martyr provides no model for continuity; June as car can finally literally come home.

The presence of the cyborg thus simultaneously erases and reinserts June into being. As Anne Balsamo notes in *Technologies of the Gendered Body,* by "reasserting a material body, the cyborg rebukes the disappearance of the body within postmodernism. Yet it never contradicts the variety of discursive constructions of the female body. The cyborg connects a discursive body with a historically material body by taking account of the way in which the body is constructed within different social and cultural formations" (33). When Erdrich mixes June and the car, she evokes the Native American automotive tradition laid out by Deloria, both challenging the stereotypical construction of the squaw by aligning her with automotive technology and recognizing the material conditions of the reservation in the ways that the characters react to June as car. By basically reincarnating her in the car, Erdrich reverses June's material disappearance, reinventing her technologically rather than spiritually. June was a failure as a literal mother; it is only as a car that she brings maternal support to Lipsha.

In *The Bingo Palace* Erdrich continues to explore the ramifications of the slippage between woman and car. June returns again, this time as a ghost, but she drives off in the blue Firebird bought with her life insurance, after leaving Lipsha bingo cards that he will eventually use to win a van. Once he sees the van displayed, he realizes, "I wouldn't want to live as long as I have coming, unless I own *the van.*" By aligning his life with a car, Lipsha, too, admits to a cyborg identity. Even though he realizes that his obsession with the van represents "a symptom of the national decline," he cannot resist the allure of the car, possibly because, unlike the Firebird, this car is *not* his mother. This van, it must be noted, is no mother's minivan; rather, it seems to be hybrid van, complete with mini-refrigerator and sleeping platform, the accoutrements of a VW 1960s microbus and a more contemporary no-holds-barred vehicle, fully wired for sound. It seems to him like a "starter home, a portable den" (63). The reference to a home evokes the minivan, but the mention of the den—generally perceived as a masculine space—makes it clear that this is a man's car. Thus, even though June's ghost provides him with the winning bingo card, there remains a separation between the mother and the machine. While the van does not exactly bring him good fortune, it does allow him to assert his own agency, as June completes her mothering by sending her son on his way.

June is not the only mother associated with cars in the text. Fleur Pillager, Lipsha's great-grandmother, the strong woman with seemingly supernatural powers portrayed in *Tracks,* returns with a young white boy and a white Pierce Arrow to regain her lands from Jewett Parker Tatro, the former Indian agent. While the community initially believes that the young white boy will be the bait to lure Tatro into a poker game, they soon realize that, in fact, the car serves that purpose. "He walked alongside the machine, stepping with the eager reverence of a prospective owner. Once or twice he smoothed his hand across the hood, kicked the tires, jiggled the grille, and tugged the chrome bumpers. . . . Jewett Parker Tatro had in his life managed with such thorough ease to acquire anything that pleased him—beaded moccasins, tobacco bags, clothing, drums, rare baskets, property of course—that when he saw the car he made an immediate assumption. He could get it from Fleur, just like he had acquired her land, and he would" (*Bingo* 143). Falling into the trap of assuming white superiority, Tatro allows himself to be blinded by the car. Erdrich subtly reverses the age-old image of taking advantage of Indians by offering shiny trinkets. Here the trinket is the car, which Fleur does not even have to offer in exchange for the land; she wins the land and keeps the car. Fleur, in some ways the spiritual mother of the entire reservation, thus exploits the car to regain the lost land, proving once again the integral role of the automobile in a Native American community—already implied by its inclusion in the list of traditional Native American objects that Tatro accumulates. Indian mothers with cars recoup some of the losses imposed by white culture. This constitutes a very different kind of maternal automotive power than that presented by the auto industry, which relegates women to driving nurseries on wheels, even if they do have torque converters and turbo engines.

Indeed, the maternity associated with the car ultimately transcends gender in this novel. With a similar ending to *Love Medicine*, Lipsha ends up driving his yet again newly escaped father, Gerry Nanapush, to freedom yet again. Crawling through a blizzard in a stolen car, they suddenly encounter the blue Firebird with June at the wheel. They follow it through the snow until Gerry jumps into the car and drives off with June. Again, Erdrich makes no distinction among human, machine, and otherworldliness: the car is real—or once was. Lipsha and Gerry are alive; June is dead. As Haraway notes, "A cyborg world might be about lived social and bodily realities in which people are not afraid of their joint kinship with animals and machines, not afraid of permanently partial identities and contradictory standpoints" (154). In this novel, in which human and machine are elided, Erdrich enhances the cyborg universe even more by adding a slippage between life and death into the mix. She presents a world of social realities in which Indians are policed by white laws yet able to escape white jails, in which Bingo Palaces and resorts provide much-needed revenue that may justify what seems to some a desecration of tribal lands, in which mothers can turn into cars, and in which mothers can use cars to reacquire land. Because Lipsha can accept all the seeming contradictions of his cyborg world, he finds peace and acceptance.

This cyborgean imagery is very different from the picture presented by Harry

Crews's wacky 1972 novel, *Car,* in which Herman Mack is determined to eat an entire Ford Maverick. After his first bite (encased in a plastic capsule) Herman dreams of cars: "His eyes filled with cars. They raced and competed in every muscle and fiber" (73). The dream turns nightmarish until, "at the last moment, when he was gasping and choking with cars . . . a solution—dreamlike and appropriate—came to him in his vision. He was a car. A superbly equipped car. He would escape because he was the thing that threatened himself" (74). Herman finds both solace and terror in his automotive reincarnation as he envisions replacing all his "parts" to gain immortality. "Replace everything with all things until he was nobody because he was everybody" (75). Herman as cyborg becomes both everyman and no-man, yet even this capacity to replicate himself infinitely falls short of the kind of transformation Lipsha undergoes in *Bingo Palace.*

Lipsha, in essence, becomes a mother. For the stolen car contains a baby. Lipsha has trouble seeing "that it's a baby because I am behind on the new equipment. He sits in something round and firm, shaped like a big football, strapped down at the chest and over the waist, held tight by a padded cushion" (*Bingo* 253). The baby's identity is nearly masked by the automotive accoutrements designed to reinforce maternal care. Yet despite all the baby technology, the child feels like a "weight" in Lipsha, "heavier than a shrunk-down star" (255). In feeling this bodily connection to the child, Lipsha duplicates pregnancy without appropriating it. He's not claiming maternal power; what he feels is maternal responsibility for the well-being of the baby. After Gerry drives off with June, Lipsha releases the child from the equipment and zips him next to his body to keep him warm, undoing the abandonment he felt as a child: "here is one child who was never left behind" (259). He also undoes automotive maternal technology, placing the child next to his body, not in the child-restraint system. In so doing, he replicates the mother-child duality, flesh to flesh. Unlike Herman Mack, who relinquishes his identity to the car—"He saw clearly that he was defined by the car, that his very reason for living was bound up in the undigested and undigestible parts of the Maverick that still had to be swallowed" (Crews 113–14)—Lipsha finds his identity expanded; he finds reasons for living in saving the baby rather than in eating a car.

Patrice Hollrah has examined the strength of Erdrich's female characters, looking at birthing scenes in which other women support and aid in the process and suggesting that Erdrich privileges these more "traditional" practices (98). But while Erdrich may support this female-enabled maternity, she also extends maternity, accommodating the Native American tradition of nurturing men as well as strong women. By using the car to collapse not just gender but also traditional concepts of maternity, Erdrich, to use Balsamo's term, "reconceptualizes" the body "not as a fixed part of nature, but as a boundary concept" (5). The maternal body functions as a boundary between woman and car, between man and woman, and between life and death. By eliding the boundary between mother and car, Erdrich offers up preternaturally powerful mothers, claiming the automotive cyborg as a powerful spirit in this Native American community and reminding us that maternal cars come in all sorts of permutations.

In some ways not much has changed over the course of the automobile's first century. We no longer call the panels around the engine "skirts," though *bonnet* remains the standard term for the car hood in Britain. The Tin Lizzie is largely extinct, but cars are still predominately gendered female. And Ford's 1912 claim that the motorcar would make a woman "more the woman" has come rather eerily true. And yet the extension of female power beyond the female body that the car can provide is also a reality. In taking on automotive power, women have moved far beyond whatever sphere may have been constructed to constrain them. While today's automakers continue to see in the female body a means for marketing their product, many women novelists are subtly challenging the meaning of that leaky distinction. Certainly, none of these writers view the slippage between woman and car as entirely positive, but they do recognize that technology has helped to transform female agency and maternity itself. In playing with the link between woman and car, they both reflect and revise our cultural understanding of women's identity and of gender itself in an age of technology.

Ultimately, by exploring the conjunction of woman and car, we can appreciate the ways that contemporary women's fiction reconfigures gender and maternity. Balsamo writes, "Cyborg identity foregrounds the constructedness of otherness. Cyborgs alert us to the way in which identity depends on notions of 'the other' that are arbitrary, shifting, and ultimately unstable" (33). Given that mothers may be seen as "othered" in their very existence—pressured to sacrifice careers, finances, and even their lives for the sake of their children, expected to live for and through their offspring—the cyborg offers not a way out but a rethinking of their situation. The cyborg mother draws attention to that "constructedness of otherness" inherent in maternity. A mother may be just as "constructed" as an automobile. And when the cyborg reflects a leakage between mothers and cars, it insists upon the presence of the maternal cyborg body. Cars, after all, are everywhere. Cultural messages regarding mothers' automotive responsibilities play on every TV screen, every magazine, virtually all parenting guidebooks and Web sites. By aligning the mother with the car, however, these writers transform maternity into something material and mobile, natural and mechanical, and even masculine and feminine. Rather than devaluing the mother, the car renders her central to automotive culture. If Ford does, in fact, listen to its mother, the entire industry requires maternal knowledge, knowledge that the auto industry may construct as juice boxes and car seats but that women writers construct as much more powerful and far-reaching. The mother as car reshapes maternity into something less essentialized, less dependent on one's biological function and more in tune with a range of cultural formations and constructions, thus destabilizing the supremacy of any of them. Maybe we didn't want to watch *My Mother the Car* because it hit too close to home; suggested the possibility of a genie without a bottle, a witch without a controlling husband. If the mother *is* the car, then one cannot drive away and leave her behind. As long as we live in a culture of automobility, we must rely on mothers and cars.

CHAPTER FOUR

Getaway Cars

Women's Road Trips

In an essay entitled "Women and Journeys: Inner and Outer" the travel writer Mary Morris asserts that "women's literature from Jane Austen to Virginia Woolf is mostly a literature about waiting, and usually waiting for love. Denied the freedom to roam outside themselves, women turned inward, into their emotions. . . . For centuries it was frowned upon for women to travel without escort, chaperone, or husband. To journey was to put one at risk not only physically but morally" (25). Her comment appropriately identifies Virginia Woolf as the endpoint of a long tradition in women's writing. Although many would contest this sweeping indictment of early women's writing as stationary—and there are surely numerous exceptions—she identifies a widespread, if not universal, characteristic. The twentieth century, however, brings a marked change. The advent of the automobile allowed women to hit the road in numbers and to do it alone. Alice Huyler Ramsey's 1909 cross-country drive took place only six years after Dr. H. Nelson Jackson and Sewall K. Crocker had completed the first transcontinental road trip in 1903. Ramsey was followed by Blanche Stuart Scott in 1910 and Anita King, the first woman actually to make the drive solo, in 1915.[1] The first woman to drive a motorcycle coast to coast was Mrs. Harry Humphries in 1913 (Pierson 149).

This development opened up not just opportunities for women but opportunities for women's literature. No longer relegated to waiting, women wrote increasingly about journeys, about mobility, and about the power inherent in this increased freedom. The motif of the journey, so long associated with men—from Odysseus to Sal Paradise in *On the Road*—came up more and more in women's texts. But if Western culture and Western literature had been predicated upon the woman in the house, then the presence of women on the road radically unsettled assumptions of domesticity, gendered identity, and gendered literature. Sidonie Smith has observed, "If traveling, being on the road, makes a man a man—and makes masculinity and its power visible—what does it make of a woman, who is at once a subject as home and a subject at home?" (x). Along the same lines Karen Lawrence asks, "How is femininity constructed when its relation to the domestic is radically altered?" (x). To take her question even further, one could ask: how are femininity and domesticity constructed on the road—and in a car?

There are many ways, of course, of being on the road. Recent theory abounds with references to nomads, migrants, travelers, refugees, and exiles. Not all movement is available to all people, however, and mobility, once thought to challenge the dominant center, is now recognized as being a situated, and often privileged, condition. And mobility itself is shaped by gender. Janet Wolff argues in a very influential essay that "just as the practices and ideologies of *actual* travel operate to exclude or pathologize women, so the use of that vocabulary as metaphor necessarily produces androcentric tendencies in theory" (224).[2] If, as Wolff claims, there is an "*intrinsic* relationship between masculinity and travel" (230), then to explore women's travel demands significant retheorizing and recontextualizing of concerns about mobility. Much work in that direction is currently taking place in a wide variety of disciplines, often concentrating on the ways that travel refigures, challenges, and revises women's relation to the domestic sphere. As feminist geographer Linda McDowell points out, "Travel, even the idea of travelling, challenges the spatial association between home and women that has been so important in structuring the social construction of femininity in the 'West,' in Western social theories and institutional practices" (206). But *travel* is a generic term; *how* one travels plays a crucial role in journey narratives and fictions, one that often gets overlooked in studies of the genre. "When women do travel," cautions Wolff, "their *mode* of negotiating the road is crucial" (233). What one moves is a body—in particular, a female body. The means through which women accomplish such a feat can spell the difference between reaching the journey's end or being raped and/or murdered along the route. The right or wrong car can make or break a road trip.[3] Women's mobility and women's agency are shaped by the vehicle that moves the female body.[4]

Agency has become a highly contested term. Does the belief that an individual can exercise independent agency relegate us to the limitations of bourgeois humanism, ignoring cultural and political forces that may shape our alleged free will? Yet if we decide that agency is constructed by a nexus of history, governmental control, and cultural discourse, does that deny personhood? The two sides in the ongoing debate seem unnecessarily far apart; indeed, its persistence indicates that agency is not easily dismissed, even if it must be qualified. Judith Kegan Gardiner, in her introduction to *Provoking Agents: Gender and Agency in Theory and Practice,* suggests that "the conviction that agency is or should be 'most one's own' is not the result of a natural essence but is a feminist belief about human fulfillment. The denial of agency is a denial of personhood that polarizes power relations." Feminists, she goes on to say, "claim that the capacity to become an agent is potentially available to all people, but that such capacities are shaped in interpersonal and discursive fields of power that may inhibit or enable them. Each person's potential for activity will also be shaped throughout life as behaviors are repressed, rewarded, learned, and transformed in the practice of organizations and institutions from the family to the state, from the university to the feminist consciousness-raising group" (13). Or, as Ellen Messer-Davidow puts it, "Agency . . . is neither a capacity of the individual nor a function of the social formation, but the co-(re)constitution of individual practices and social processes" (30). Women drivers exercise a particular form of agency, granting them the freedom of independent movement, yet they remain subject to state regulations, financial status, domestic obligations, and simple opportunity. Thus, I use *agency* to denote the exercise of actions that stem from individual choice, though that choice is mediated by social, political, and, particularly, domestic forces. In examining women's car trips, one often finds that women's mobility models women's agency as a combination of individual action and larger social and economic forces.

Automobile journeys integrate the road and domesticity in ways that highlight the extent to which women on road trips are situated, bound by gender, by economics, and by a web of relations. Whatever agency is exercised on the road is necessarily limited and constrained; one might even say that women on the road are at least partially immobilized. Given the problematic nature of mobility, this sense of being tied down while mobile serves as a key component in challenging the claim that to be on the move is necessarily empowering and broadening. As Elizabeth A. Pritchard asks, "Why would feminists enjoin mobility, per se, as the measure of progress? If it would be passing strange that a story of women's development tell of our containment, why is it not just as jarring to have *dislocation* be the story of women's development?" (45–46). The car, however, mitigates—without necessarily erasing—such dislocation; women in cars have the privilege of taking a kind of home with them on the road. Neither fully contained nor fully mobile, women in cars call into question both domesticity and movement as empowering female tropes and, more particularly, as mutually exclusive spheres. Car trips highlight the dangers inherent in an unqualified celebration of mobility as a means of escaping the centering of power. Certainly, the 1991 movie *Thelma and Louise* made clear that women on the road do not escape the policing power of the state. Some forms of mobility are more powerful than others.

The "discourse of mobility," Pritchard argues, "is itself an axis of power" (53). It is a privilege to be able to hit the road, either literally or theoretically. Road trips undertaken in contemporary women's fiction, however, reveal the uneasy contradictions underlying women's mobility: the possibility of freedom and the constraints of domesticity, the ability to escape the physical confines of the home and the reconfiguration of home as mobile. As the genre develops, we see more daring adventures, trips that leave domesticity behind altogether and overtly challenge the classic male road narrative. And yet one thing remains constant: women, just like men, need money to operate a vehicle.[5] As so many mobility theorists have pointed out, mobility is not free—in any sense of the word. Wolff reminds us that the "suggestion of free and equal mobility is itself a deception, since we don't all have the same access to the road" (235).

Mary Louise Pratt, writing on travel narratives by both men and women, notes: "The predictable fact that domestic settings have a much more prominent presence in the women's travel accounts than in the men's . . . is a matter not just of differing spheres of interest or expertise . . . but of modes of constituting knowledge and subjectivity. If the men's job was to collect and possess everything else, these women travelers sought first and foremost to collect and possess themselves. Their territorial claim was to private space, a personal, room-sized empire" (159–60).[6] Others have pointed out that Western women traveling abroad generally benefited strongly from their colonialist positions, despite their beliefs to the contrary;[7] even Pratt recognizes that a room of one's own can constitute an empire. But her point, that women on the road follow different modes of "constituting knowledge and subjectivity," is an important one. Traveling by car forces women into a constant negotiation among home, family, and the road, highlighted by the increasing need for technological sophistication as both cars and road trips become more complex. Mobility, never easy, is both enabled and further complicated by the demands of automobility.

If domesticity goes on the road, not only are new domestic skills needed, but, as Pratt indicates, so are new modes of constituting knowledge and subjectivity, especially given the ways that the mobility can transform traditional means of determining female identity as linked to home. To be a woman on the road requires, first of all, some knowledge of cars; this in itself challenges the stereotype of women as mechanically inept and dependent. In *The Bad Girl's Guide to the Open Road* (1999) Cameron Tuttle hypes women's road trips as opportunities to eat greasy food and exhibit outrageous behavior. Yet she also spends considerable time detailing some of the most common car problems along with advice on how to fix them. A road trip, she insists, hinges upon the connection between a woman and her car: "The first step to badness is understanding the relationship between you and your car. If you're like most women, you've been conditioned to think of a car as a mobile chatroom, or a big purse on wheels, or even a high-speed motorized shopping cart. . . . But a car is much more than that—it's your freedom fighter, your power booster, your ticket to ride. It's a stimulant, an antidepressant, and a vroom with a view" (48). Once one learns to understand and celebrate what the car offers, road trips provide access to a whole new public sphere. Women in cars, forced to develop new kinds of expertise, gain an increased awareness of their cultural situatedness and women's place in a technologically driven world. In order to explore these issues through women's road novels, one must also take into consideration how the presence of women on the road transforms the narrative of the road. Women on the road may unsettle gender and domesticity; women who write about being on the road challenge the form itself.

The latter part of the twentieth century saw a remarkable rise in the number of women who write about cars. From novels to auto care guides car culture increasingly targeted women. So did the auto industry. J. D. Power and Associates, the major automotive research firm, had begun studying women in earnest, according to a 1982 piece in *Automotive Age.* In the 1980s, according to Andi Young, the auto companies varied in their appeals to women, some (Ford and Chrysler) continuing to single them out and others (GM) deciding that "women's car desires are essentially the same as their male counterparts" (Young 11). A Chevrolet publication in 1986 provided tables and questionnaires to car dealers and salespeople, informing them of the tremendous impact that women have in car buying: women, according to "Pretty Soon Every Other Guy Who Walks into Your Showroom Will Be a Woman," purchase 44 percent of all new vehicles and influence 87 percent of all car purchases. "You can't live without them," the front cover wryly announced.

All of this indicates that women, by the 1980s, had a recognized place in car culture. Yet that place was far from equal to that of most men. Lyn St. James, one of the pioneering female race car drivers, noted in 1993 that auto companies had still not succeeded in learning how to deal with women, arguing that most women still felt talked down to by car salesmen. And while the auto industry has recognized the tremendous economic potential of women as consumers, other cultural voices have resisted women as drivers. Beth Kraig, in "The Liberated Lady Driver" (1987), cites a 1984 article by Robert Weidner that claims the "vast majority of American women regard the automobile as something: (1) that you occupy between Where You Are and Where You Need To Be; (2) that is likely to turn on you at any moment; and (3) that makes a reasonably functional (albeit mobile) combination closet and vanity" (qtd. in Kraig 379). It is assumed, says Kraig, that women "domesticate the car" (380). She goes on to elaborate the stereotype that the "woman's car may remind her of the home to which she will gladly soon return; the man's car can attest to the fact that the road *is* his home" (382). Despite the sophistication of marketing cars to women, then, cultural assumptions about women and cars still linked women's cars to domesticity, as an extension of the home.

Women, however, challenged such assumptions. As an article in *Autoweek* in 1983 put it, "I don't know if they've ever addressed this particular aspect of womanhood, the Friedans, the Millets and Steinems, the Rita Mae Browns, but I suspect one of the most insidious elements of sexism is Feminine Car Denial" (Carlson 8). Relegating women to boring family cars, claims Satch Carlson, denies them fulfillment. By linking cars to the escalating feminist movement of the 1980s, Carlson makes a very important move. Cars, after all, have a history with the women's movement. In 1910 the Illinois Equal Suffrage Association embarked on a series of road trips to promote the cause. "From this time forward, the movement would use the auto not only as a convenient form of transportation," says historian Virginia Scharff, "but as public platform, object for ritual decoration, and emblem of the cause of women's emancipation" (79).[8] The car continues to be an emblem of women's emancipation. Automotive journalist Lesley Hazleton, based on a series of focus group studies in the early 1990s, concludes that "while men take for granted the independence that cars bring, women do not. Our own car means freedom. It means control of our own lives" (*Everything* 3).

Women novelists twist these strands of optimism and independence, however, recognizing what may be entailed in the economic independence and what dominant forces may be beyond the car's power to alleviate. Marie T. Farr, examining automobiles in women's fiction of the 1970s and 1980s, anticipates Hazleton, arguing, "The automobile has matured into a symbol for women's independence, self-mastery, and even creativity" (168). This is certainly true, though cars and mobility do not always appear as emblems of unqualified and celebratory female agency; when women domesticate the automobile, they find both empowerment and constraint, often forced to acknowledge that technology and car culture are still male dominated and that money matters, as do race, ethnicity, age, sexual orientation, and physical ability. Cars do not liberate women from home or domesticity. But by eliding the boundaries between car and home, they do open up the possibility of reconfiguring women's place as both situated and mobile, both domestic and independent.

Bobbie Ann Mason's *In Country* (1985) and Barbara Kingsolver's *Bean Trees* (1988) illustrate the perils and possibilities for women on the road. Mason's Sam Hughes and Kingsolver's Taylor Greer experience the questionable mobility and empowerment provided by ancient and unreliable VW Beetles. In each novel a limited form of women's agency is facilitated by the automobile, but these road trips do not offer the kind of transcendental "IT" sought by Kerouac and his ilk. As Sal Paradise puts it in *On the Road* (1957), "We all realized we were leaving confusion and nonsense behind and performing our one and noble function of the time, *move*" (Kerouac 133). Through the course of the narrative Sal may come to realize that movement is less "noble" than he initially believes, but he, unlike Sam or Taylor, is given the chance to assert the claim. Women in these narratives do not escape attachments, domesticity, or responsibility. They cannot head out wherever their fancy takes them, with a blithe disregard for money or family, as does Dean Moriarty, leaving wives and children (four by Sal's last count) behind. They do, however, significantly revise the old associations of woman as home, woman as place. In fact, they revise the concept of home as space, refiguring both female subjectivity and domesticity. Lawrence, examining British women's travel writing, notes that women writers "replace the static mapping of women as space . . . with a more dynamic model of woman as agent, as self-mover" (18).

Women in cars, however, are not quite "self-movers"; rather, they are *drivers*, women who move themselves, their home space, and a machine to boot. By putting women at the wheel of unreliable cars, Mason and Kingsolver challenge not only women's place but also the genre of the road trip itself. In their decrepit clunkers Sam and Taylor drive beneath the radar of the automobile industry that so actively seeks their purchasing dollars for new cars. From the start, then, these characters construct an alternative to industry-defined car culture, resisting the assumptions of what women want in a car. While this position aligns them with the figures in Kerouac's *On the Road*, who largely drive old cars, generally owned by other people, these women find themselves confronted more directly with the mechanical difficulties of living on the margins of car culture, thus reminding them yet again of their vulnerability in the midst of the thrill of mobility. Scant financial means can limit one's comfort and safety on the road, if not one's ability to hit the road.

The female road novel, however, takes a new turn in Erika Lopez's 1997 *Flaming Iguanas: An Illustrated All-Girl Road Novel Thing*, in which Tomato Rodriguez, narrator and protagonist, experiences heightened thrills and vulnerability on her cross-country motorcycle journey. Barely able to drive a motorcycle, let alone to repair it, Tomato pushes the boundaries of the road novel in many ways, including replacing the car with the motorcycle, a vehicle even more strongly associated with masculinity and exposing its rider to even more danger. Barbara Joans, writing about women and the culture of Harley-Davidson motorcycles, claims that women bikers are perceived as "gender traitors" by the culture at large (89). By claiming this classically male space, Tomato, a self-proclaimed "biker" and "biker chick," collapses the embattled boundaries of masculine and feminine spheres *and* masculinity and femininity. She constructs an anti-domestic female and lesbian identity, as she finds a kind of mobility unavailable to Sam or Taylor, and opens up the road narrative to a wider range of female possibility, while still recognizing its limitations.

Cars, War, and Pop Culture

Mason's novel *In Country* examines the possibilities of liberation, mobility, and selfhood that the automobile offers. The spirited protagonist may achieve a sense of fulfillment, but the car's power to provide answers—or even escape—is extremely limited. Sam invests a great deal of emotional capital in her rickety Volkswagen Bug, but she hardly experiences any kind of "free" humanist autonomy or agency while on the road, where her power remains as tenuous as the rusted car with a faulty transmission. With the increasingly sophisticated development of the automobile, most people, particularly women, are at the mercy of the machinery; the ability to fix one's car on the road is increasingly rare. The VW Beetle may be the later-twentieth-century equivalent of the Model T, with its relatively simple engineering, yet the idea of actually fixing the car herself seems a remote possibility to Sam. This alienation from the machinery—found in both men's and women's literature—instigates a retreat from the excitement of technology; the automobile is now a fact of American culture rather than a striking new innovation.

While girls and boys books of the 1910s highlight the challenge of running a car, and the economic survival of the Joad family in the 1930s depends upon their keeping the car in working order in John Steinbeck's 1939 novel *The Grapes of Wrath,* in the second half of the twentieth century the emphasis seems to move to the journey rather than the vehicle. Sal and Dean of *On the Road* care little about what car they drive or the intricacies of auto mechanics so long as they keep moving. When Dean arrives in a new Hudson, Sal notes that neither the heater nor the radio works: "It was a brand-new car bought only five days ago, and already it was broken" (Kerouac 116). As long as he can keep the engine running, Dean has no concern for the car as a working machine. Yet issues of danger and domesticity keep the vehicle at the forefront of women's texts; when one is perceived as more vulnerable to assault, a reliable car becomes even more important, also a critical consideration when one has one's children in the car. Thus, the car often functions as the site that calls into question the issues of female agency, female power, and gender itself; how can one be a woman and/or a mother in the car? Going on the road challenges what it means to be a woman and where her place should be.

Sam's angst is set within a uniquely American background. The novel takes place during "the summer of the Michael Jackson *Victory* tour and the Bruce Springsteen *Born in the U.S.A* tour" (23). The context is defined not just by singers but by tours, highlighting the extent to which American culture has mobilized. In this summer of 1984 Sam decides that she suffers from the post-traumatic stress syndrome of the

Vietnam War, as she tries to learn about the father killed there a few months before her birth. In her attempts to come to terms with her father and with her invented nostalgia for the 1960s, Sam finds that the confines of Hopewell, her small Kentucky town, severely limit her search to sort out who she is and what she wants. In the early part of the narrative her complaints center on her enforced immobility and the expectations that go along with being female and car-less: "Boys got cars for graduation, but girls usually had to buy their own cars because they were expected to get married—to guys with cars" (58). Cars, then, mark not only gender—boys are entitled to them—but also dominance. Keeping women without cars pushes them into marriage and thus rootedness and domesticity, reinforcing the traditional association between woman and home. As feminist geographers Monu Domosh and Joni Seager note: "The control of women's movement has long preoccupied governments, families, households, and individual men. It is hard to maintain patriarchal control over women if they have unfettered freedom of movement through space" (115–16). The "patriarchal grip slips," they go on to say, "when women get cars of their own" (121). This is precisely Sam's intention; part of the appeal of a car is to escape marriage and motherhood, to avoid domesticity.

Sam's desire for a car functions at many levels, both superficial and profound. As she recognizes, it would provide access to distant malls, bars, and fast food restaurants, the staples of American teenage culture. But Sam also seems aware that her longing for a car is for more than adolescent shopping trips. Living with her uncle—a maladjusted, seemingly perennially adolescent Vietnam vet—she, despite her youth, takes on adult responsibilities such as working, shopping, and worrying, while he watches TV, wears skirts in imitation of *M*A*S*H*'s Corporal Klinger, and plays video games. Her lack of a car, however, constantly reminds her of her dependent status and reflects her frustration at the adult world that refuses to take her seriously: to tell her about the war, to tell her about her father, to tell her about the 1960s. As long as she is car-less, she cannot take off on her own to find some answers, and the adults around her repeatedly deflect her questions. In fact, they tell her that since she's a runner, generally logging in six miles a day, she shouldn't need a car. A six-mile radius should be enough range for any young woman; she should stick to running on her own power rather than muscling her way into automotive power. But Sam realizes that to rely only on the body in this age of technology mires her in small-town America, and even in the age of TV she finds this confinement stultifying.

A car represents for her the possibility for independence and a true adult life. These desires all come together, predictably, when she meets Tom, a Vietnam vet twenty years her senior with a car to sell. Although neither Tom's 1973 VW beetle with a faulty transmission nor Tom himself, impotent as a result of psychological trauma in the war, seem strong candidates for sexual charisma, Sam fixates on both and buys the VW with some help from her mother, who hopes that providing Sam with a car will enable her to drive to Lexington, where her mother lives with her new husband, thereby reestablishing the mother-daughter bond. For Irene, Sam's mother, the car will thus reinforce family ties. But Sam's primary response to the automobile is feeling Tom's presence "everywhere in the car" (6). Despite her disappointment in Tom's lack of sexual ability and the car's transmission, she immediately realizes that "owning a car gave her power" (176). While the car initially serves as an emblem of sexual desire, it rapidly becomes the means for Sam to begin enacting pieces of her father's past.

It is significant that the sexual metaphor embodied in this vehicle is that of failure; neither Tom nor the car perform very well. She is beginning to grasp Hazleton's blunt claim about the power of the car: "Women know where the real power is. We know that engine size is as irrelevant as penis size. Power is not in the numbers, or even in the performance, but in the sense of control and independence" (*Everything* 21). While Sam's attraction to Tom leads her to buy this particular car, it very soon becomes more than a sex symbol;[9] it becomes a means to create a past and, even more, to create an experience denied to most women: the experience of war. She drives to the swamp in an attempt to recreate what it means to be "in country." More significantly, she, along with her uncle and grandmother, sets off for Washington, D.C., to visit the Vietnam War Memorial. It is the trip there, as much as the sight of the wall, that moves Sam closer to a kind of peace.

The road trip, broken up by stops at Exxon stations, Howard Johnsons, and Holiday Inns, is set in a quintessentially American landscape and therefore seems appropriate as a setting to try to resolve the very American problem of the legacy of Vietnam. As Sam realizes, "Everything in America is going on here, on the road." But what she means by that is unclear. Largely, it appears that the sight of traffic headed in all directions invigorates her: "They are at a crossroads: the interstate with traffic headed east and west, and the state road with north-south traffic. She's in limbo, stationed right in the center of this enormous amount of energy" (17). Sam positions herself as central, the standard location for those in the so-called First World. While this assumption certainly marks her privilege, that her excitement is generated not so much by her position as by her proximity to moving traffic mitigates somewhat the hierarchy implicit in her central location. Caren Kaplan has cautioned that the First World feminist critic needs "to examine her location in the dynamic of centers and margins" ("Deterritorializations" 189). It seems slightly unfair to saddle this fictional character with the obligations of a cultural critic, particularly given her age; besides, Sam's centrality is constructed only as the nexus around which movement happens. She finds meaning not in being grounded while cars around her move but in the possibility of joining any of those streams of traffic. This does not erase her relative privilege, but it does resituate women's place as on the road rather than in the home. The potential for movement—in any direction—offers a feeling of the real that living in a small town does not: "On the road, everything seems more real than it has ever been" (7). And yet that reality consists of still more evidence of mass culture: chain restaurants, service stations, and hotels.

Ronald Primeau has suggested that the "primary motivating force of the American road quest is a longing to return to the time when the stream was deeper, when local customs and regional culture were preserved—a time before the landscape was pasted over with billboards and interstates circumvented the old highways" (65). Like Edith Wharton, he sees the car as means of rediscovering the past. Sam's road trip, however, while it seeks to recover the past of a dead father and a lost war, revels in the billboards and interstates, for that is how so many people, particularly adolescents, define themselves in contemporary American culture. Louise Erdrich notes that as our culture becomes more mobile and we lose a sense of place, "there still remains the problem of identity and reference." Thus, whether "we like it or not, we are bound together by that which may be cheapest and ugliest in our culture, but which may also have an austere and resonant beauty in its economy of meaning. We are united by mass culture to the brand names of objects, to symbols like the golden arches, to stories of folk heroes like Ted Turner and Colonel Sanders, to entrepreneurs of comforts that cater to our mobility, like Conrad Hilton and Leona Helmsley. These symbols and heroes may annoy us, or comfort us, but when we encounter them in literature, at the very least, they give us context" (Erdrich, "Where" 46). Our context for understanding how Sam positions herself lies in her connection to popular culture; for example, she feels the death of Colonel Henry Blake on TV's *M*A*S*H* more acutely than that of her own father. Just as her excitement in the movement around her ungrounds her, so the anonymity of consumer culture facilitates her inclusion in the American cultural experience.

Thus, the road reassures her that all places are alike and that she is finding different experiences, both similarity and difference. She loves the hotel rooms because they show "no evidence of belonging to anybody, but [they have] a secret history of thousands of people, their vibrations and essences soaked in the walls and rugs" (12). Cultural critic Meaghan Morris has examined the import of hotels in travel narratives and travel theory. While hotels, as she notes, can provide safety for women travelers and serve as a "transit-place" (2), they can also "*demolish* sense-regimes of place, locale, and 'history.' They memorialize only movement, speed, and perpetual circulation" (3). Such anonymity is precisely what Sam seeks; she might be anywhere, finding purpose only in perpetual movement. "I wish I'd wake up," she says en route, "and not know where I was" (6). In preferring anonymous hotel rooms, belonging to no one, to the conventional family home, Sam reflects her resistance to traditional domesticity. She needs to be both unfettered, belonging to no one, and connected, feeling the "vibrations" of others. She thus resists the construction of a bourgeois humanist self, opening the door to an identity based on positionality and movement. The car provides the ideal trope for such a self, functioning both as an individual vehicle and a part of traffic.

Driving her own car through the American countryside to the American capital city, Sam seems to gain a sense of belonging that has eluded her until this point. Allowing her to be in control of her direction and destination, the trip moves her from adolescence to adulthood, as she comes to the wall and finds not only her father's name but also her own—a ghostly extension of herself as a literal casualty of the war. The death of a different Sam Hughes conveys to her a further connection to the war no one will talk about. This car trip ends, then, with tears and reconciliation with her dead father, seemingly a successful journey. But the cost—literally—has been high. The car, bought by a teenager infatuated with an older man, breaks down on the way and is repaired only because Sam has her mother's credit card for emergencies. She may control the wheel of the car, but its performance is underwritten by her mother and stepfather.

Tellingly, driving constitutes only one element of the forces needed to complete the journey; economics is even more critical, and Sam has little independent control over money. Thus, while one may celebrate the agency that leads her to the monument, one also has to realize how much of the trip is controlled by other forces. To use Pritchard's terms, this is hardly a narrative of development through mobility; the focus on stops, hotels, and consumer culture muddles the linear shape of the classic road narrative. This results in a more nuanced road trip that reflects the fragility of a girl's place in car culture; the car enables a kind of agency, but the trip reminds us of the many contexts in which individual action and will are simply ineffectual, the naive beliefs of a teenager that having a car means she is in charge. The women of Hazleton's studies may express unqualified enthusiasm for the control a car offers, but Mason, looking at a larger context, acknowledges that cars only take you so far—not to the metaphorical journey's end of full control and agency.

In fact, the novel ultimately confronts more than a girl's search for her dead father and a bygone era that she only imperfectly understands. It also poses the question of what constitutes identity in a culture in which so many people are constantly on the move and each place begins to look more and more the same: the same malls, the same restaurants, the same TV shows, the same problems. The automobile reflects this dilemma by providing the individual freedom (though monetary constraints still hold) to set forth on an identity quest only to realize that the places one sees are not much different from the town one left behind. It thus provides the aura of autonomy while still imposing conformity to mass culture. But Sam is not just seeking an American identity and an understanding of the American legacy of the Vietnam War. She is also questioning what it means to be a woman in a technologically driven country where technology seems to be the province of men. She is dependent on Tom's assurances that he has fixed the transmission and then tries to claim more knowledge than she apparently has when the transmission goes out, using the jargon of auto repair to deflect criticism. "I know about frozen transmissions" (17), she says in response to her uncle Emmett's challenge to her mechanical acumen. It seems fitting that of all the car problems Mason could have selected, she astutely picks the transmission to serve as the culprit. This text also lays bare faulty transmission of history as well as the failure of consumer culture to transmit substantive possibilities for identity.

Despite her plans to paint the car black and acquire a black leather jacket and black boots, conveying an image of toughness and power, Sam seems to realize that this rickety Bug offers only a tenuous foothold into the masculine world of tools and technology. Women in cars, as Scharff points out, may have "challenged the social limits of femininity" at the beginning of the century (67), but automobile culture has found new ways of reminding women of their limited access to knowledge; women, it is assumed, know nothing about cars other than how to drive them. Once driving becomes nearly universal, mechanical ability succeeds it as the test of true mastery. That Emmett knows as little about cars and transmissions as Sam is immaterial; it is her ignorance that undercuts her pride and automotive power. They are indebted to a male mechanic and her mother's credit card (made possible by the affluence of her new husband) for the continuation of their journey. One needs either automotive expertise or money to harness the power of automobility. To quote Henry Ford again, "We are *not* living in a machine age, *we are living in the power age.*" The car—as Ford knew well—is more than a mere machine, meaning that the ability to control it takes on great resonance. Sam's failure, then, does not just deprive her of transportation; it reflects her exclusion from both power and machines.

The car's breakdown confirms what Sam has already confronted as the unassailable wall of male privilege: the experience of war. Horrified by reading her father's war diary, with its claims of gleeful triumph at killing the enemy, she tries to comprehend "what would make people want to kill" (208). "Men," she realizes, "were nostalgic about killing. It aroused something in them" (209). Determined to understand, she spends a night in the swamp as the closest approximation to Vietnam she can find: "If men went to war for women, and for unborn generations, then she was going to find out what they went through. Sam didn't think the women or the unborn babies had any say in it" (208). The experience, however, doesn't quite work, as she realizes that she doesn't know how to dig a foxhole and that "this nature preserve in a protected corner of Kentucky wasn't like Vietnam at all" (214). She does gain some insight into fear as she hears sounds and imagines a "V.C. rapist-terrorist" and so decides to use the sharp edge of a can of smoked oysters "to cut his eyes out" (217, 218).

But Sam cannot consciously make the connection between her own willingness to commit violence and her father's war actions. Although she imagines her position in the swamp as "her father's place" (217), she rapidly learns it is no place for her. Her attempts to duplicate his experience fail ignominiously when this rapist-terrorist turns out to be Emmett, who tells her: "You think you can go through what we went through out in the jungle. But you can't. This place is scary, and things can happen to you, but it's not the same as having snipers and mortar fire and shells and people shooting at you from behind bushes. What have you got to be afraid of? You're afraid somebody'll look at you the wrong way. You're afraid your mama's going to make you go to school in Lexington. Big deal" (220). Remembering Pritchard's concern regarding the use of dislocation and mobility as a trope of female empowerment, we can see that for Sam leaving the house on this trip does not propel her on the road to self-development. This dislocation only reminds her that going on the road does not necessarily lead to positive change and can, in fact, lead to rape and terror. That Sam herself escapes such a fate does not erase its imminent threat.

The accuracy of Emmett's description of the horrors of war notwithstanding, his brutal denial of her attempts to understand underscores the gap between male and female experience. She might as well go home. Sam, who tries to imagine women and war through female issues of abortion and rape, is told she can never enter into the masculine sphere, a lesson that is reinforced by her car troubles on the road. In fact, the writer Tom Wolfe suggests that the car can replace war as an outlet for aggression as he compares the demolition derby to Roman gladiators in *The Kandy-Kolored Tangerine-Flake Streamline Baby:* "As hand-to-hand combat has gradually disappeared from our civilization, even in wartime, and competition has become more and more sophisticated and abstract, Americans have turned to the automobile to satisfy their love of direct aggression" (33). One of the vets Sam meets links car culture even more specifically to the war, claiming that paved roads would have allowed a victory: "We should have paved the Ho Chi Minh Trail and made a four-lane interstate out of it. We could have seen where Charlie was hiding, and we would have been ready for him. With an interstate, you always know where you're going" (Mason 134). With interstate highways war becomes just another road trip. Thus, it seems as if the car may be Sam's vehicle finally to understanding the war. But once again, gender gets in the way, for she is neither a soldier nor part of the male drag racing scene to which Wolfe refers;

all she can manage in the way of direct aggression is honking at other drivers and planning on painting a nearly defunct automotive relic black. She may try to take on a masculine aggressive role, trusting in the power of the car to provide her with the means, but car culture has already reminded her that she is a young woman in a world where cars and wars are controlled by men and where mobility can take you not just to a greater degree of self-knowledge but also to war.

Yet Mason, in constructing a situation in which Sam is essentially driving to Vietnam is, in some ways, offering Sam entrance to that sphere. By bringing the war home—where one can access it and replicate it by car (e.g., by driving to the swamp or to the Vietnam Memorial)—she domesticates the experience of war. As the technology that developed the car has facilitated mobility, so that same technology has moved Vietnam to America. Once the car punctures the domestic sphere, it opens the door for war to drive in. Sam manages to forge a connection between past and present at the Vietnam Wall, where she realizes that "she is just beginning to understand. And she will never really know what happened to all these men in the war" (240). Recognizing that understanding may compensate for lack of knowledge helps to ease Sam's exclusion from Vietnam. It will not, however, help her to fix her car. And we are reminded of this fact when the moment is presented through the imagery of the automobile. She watches some workmen clean the wall "like *men* polishing their cars" (242; my emph.). The familiarity of the task brings the wall home to her, but the simile also reminds us of the reality of gender in trying to bring together war, automobility, and identity. When it comes to war and cars, men are in charge. Sam achieves a moment of reconciliation at the memorial, but it occurs partly through a suspension of female identity: she finds her own name on the wall and passes, for a moment, into a masculine self: "She touches her own name. How odd it feels, as though all the names in America have been used to decorate this wall" (245). All the names in America, as long as they are men's names. Ultimately, this disturbing erasure of a female self seems tied to the emblem of the automobile, with its promise of power and control. Part of Sam's desire for a car can be traced to dissatisfaction with the limits imposed on her as a woman. It remains to be seen if puncturing that masculine sphere will re-gender it.

The image of men polishing their cars is not the only simile Mason employs at the Vietnam Wall. Sam herself "doesn't understand what she is feeling, but it is something so strong, it is like a tornado moving in her, something massive and overpowering. It feels like giving birth to this wall" (240). Initially, this line reclaims the Vietnam experience for women, who give birth to the memorial. But it also reinscribes conventional gender roles, relegating women to their age-old role of mother. Further, pregnancy has hardly been presented in a positive light in this book. When Sam's friend

Dawn gets pregnant, Sam notes that pregnancy "ruined your life" (103). The creativity Sam admires is not that of childbirth but of auto mechanics. When she visits Tom's garage, "the clutter of auto parts and empty motor-oil cans and unidentifiable, useless-looking objects fascinated her. Mechanics could recreate beautiful cars out of a mess of greasy metal pieces and rusty pipes and screws, the way a magician would mix watches and eggs in a bowl and pull out a rubber chicken" (76). Automotive creativity is both beautiful and magical, unlike pregnancy, the fate she actively seeks to avoid. And yet, other than envisioning painting her rusty Beetle black, she sees no place for herself amid such creation. Much of car culture remains out of bounds to women. Just as, according to Wolff, metaphors of mobility are inherently masculine, auto-mobility is largely androcentric as well. Given the role cars play in a culture that privileges technological mastery, they can remind women of their place as much as they may expand and refigure that place. Mason uses the car to transgress gender boundaries, but that transgression only illustrates the ways such boundaries get reinscribed by war and automobiles.

Transit Domesticity

Barbara Kingsolver unsettles the masculine boundaries of automobile culture in *The Bean Trees*. Taylor Greer, like Sam Hughes, finds herself trapped in a small Kentucky town. Like Sam, she acquires a VW Beetle, this one in even worse shape than Sam's heap, having "no windows to speak of, and no back seat and no starter" (10). Undeterred, she learns to push-start the car, and aided by her mother, who insists, "If you're going to have you an old car you're going to know how to drive an old car" (11), she also begins conquering a fear of exploding tires (she once saw someone blown over the top of a Standard Oil sign by an exploding tractor tire) by learning how to change and repair them. She soon realizes, however, that "there were some things I hadn't considered. Mama taught me well about tires, and many other things besides, but I knew nothing of rocker arms" (12). While this lack of knowledge seems to put her into a predicament similar to Sam's, Taylor asserts a stronger sense of control when her rocker arm breaks. The repair may cost her nearly half her money, but it is, nonetheless, her money rather than her mother's credit card. Further, she puts herself into the position of potential repairwoman, saying, "I should have been able to fix it myself" (13). Instead of trying to claim expertise she doesn't have (e.g., "I know about frozen transmissions"), she leaves open the opportunity to gain it. Further, by crediting her mother's teaching, especially about tires, she sets up auto repair as female expertise, handed down through generations of women.

Taylor's road trip is much more open-ended than Sam's, with no particular goal other than to drive west as far as the car will take her. But she does have a purpose in mind—specifically, to rename herself: "I decided to let the gas tank decide. Wherever it ran out, I'd look for a sign" (11). Yet rather than surrender this symbolic rebirthing process entirely to the car, she also takes control: "I came pretty close to being named after Homer, Illinois, but kept pushing it. I kept my fingers crossed through Sidney, Sadorus, Cerro Gordo, Decatur, and Blue Mound, and coasted into Taylorville on the fumes. And so I am Taylor Greer. I suppose you could say I had some part in choosing this name, but there was enough of destiny in it to satisfy me" (12). She demonstrates the same agency in choosing a destination; after having decided to go as far as the car would run, she again intervenes: "Whether my car conked out or not, I made up my mind to live in Arizona" (35). Thus, the car, rather than functioning entirely as domestic space or controlling agent, merely facilitates Taylor's own exertion of will in renaming and relocating herself. By acknowledging that a range of forces may produce our actions, Taylor (and Kingsolver) exerts agency and recognizes its complex construction amid shifting "discursive fields of power," to use Gardiner's term.

Unlike Sam, who seeks to drive into her past, Taylor looks to drive out of it: "I intended to drive out of Pittman County one day and never look back, except maybe for Mama" (10). In making this exception, she acknowledges that the car can break neither the maternal bond nor the past, that some things are more powerful than mobility. She may change her name and place, but she retains a strong sense of who she is and where she comes from. Indeed, her journey into the future encompasses a great deal of symbolic attachment to the past. While she takes off in her own wheels, she drives a car that contains few of the advances of twentieth-century automobile technology, such as windows and starters. In push-starting her car, she evokes the days of the crank engines, aligning herself with the intrepid women such as Alice Ramsey who refused to let the necessary physical exertion keep them from the automobile. Additionally, the car reveals to her cultural markings of the past—the lands of the Cherokee Nation, a Pioneer Woman Museum, and a park associated with dinosaurs. Taylor's experience with cars takes her on a drive through history: from dinosaurs to Indians to pioneers. This connection to the past is critical in refiguring the road trip; whereas Kerouac's characters seek to outrun the future while Sam is haunted by the past, Taylor recognizes that one's life cannot be erased and reconstructed at will, even if you do change your name. Family ties and cultural history intersect with individual decisions to shape identity and agency.

Taylor's trip takes her to a new vision of the future by introducing her to a woman who fixes cars, Mattie, the proprietor of "Jesus Is Lord Used Tires." Unlike Flannery O'Connor, however, Kingsolver situates the automobile firmly in the secular world; cars are linked not to grace but to female achievement. Meeting Mattie allows Taylor a greater realization of just what women can accomplish. "I had never seen a woman with this kind of know-how," she says. "It made me feel proud, somehow" (43). Mattie's automotive expertise—even greater than her mother's—helps to provide a bridge from past to present, as she fixes Taylor's car, thus enabling Taylor to rejoin, a bit more reliably, the contemporary automotive age, and serves as a model for how women can break out of past patterns of helplessness and dependence. Mattie's tire store and garage also offer a variation on the car art that Sam so admired at Tom's place: the mechanic's ability to re-create beautiful cars. Mattie, however, has created a car garden. "Outside was a bright, wild wonderland of flowers and vegetables and auto parts. Heads of cabbage and lettuce sprouted out of old tires. An entire rusted-out Thunderbird, minus the wheels, had nasturtiums blooming out of the windows. . . . A kind of teepee frame made of CB antennas was all overgrown with cherry-tomato vines" (45–46). This recycling of car parts into a garden presents a very different picture of the role of technology in the pastoral world envisioned by Leo Marx's classic work *The Machine in the Garden*. Rather than functioning as a dark reminder of the ways technology has infiltrated the pastoral scene, Mattie's garden suggests that cars can enhance the natural landscape. By combining automobiles and plant life, this garden also transcends conventional gendered expectations about technology and nature. Like women on the road, cars in the garden no longer seem incompatible. In thus establishing women's automotive expertise and refiguring the machine in the garden, Mattie challenges women's exclusion from car culture and transforms that culture so that women need no longer suspend female identity to achieve car mastery.

But despite its utopian aura, Mattie's place is no Eden. It is solidly grounded in a specific historical context. She also runs a sanctuary for political refugees from Guatemala, providing a means to save people from their pasts. Kingsolver provides a slightly revised version of the imagery Mason employed in likening the cleaning of the Vietnam Memorial to a man polishing his car. For Kingsolver auto care is associated with a woman who not only fixes broken tires and broken cars but attempts to fix broken people as well. It may seem a small detail, but the re-gendering of automobile symbolism marks a significant shift in cultural assumptions about women and cars, as car maintenance becomes part of the female experience, though that experience still maintains women's conventional position as caregivers. Yet by eliding auto care and person care, Kingsolver tweaks women's traditional roles without eradicating them. Cars may help to reshape women's lives, but they do not radically transform them. The refugee subplot foregrounds Kingsolver's awareness of the many forms of mobility and unsettles what could otherwise be read as a sentimental tale of female empowerment. Her protagonist, Taylor, may be on the road to better things, but mobility is not always positive. In this presentation of political exile Kingsolver avoids an uncomplicated presentation of the road trip as the path to maturity and self-awareness that complements her realization of the limited power of automobility for women. Taking on an even more daunting task than Mason, who explores a past conflict, Kingsolver forces us to confront the political climate of the present: the situation in Guatemala in the 1980s. Politics and cars are just as complexly aligned as women and cars.

Just as Sam realizes that she can never truly know what happened in Vietnam and that while a car trip to the Wall offers a moment of healing, it cannot truly erase the scars of the war, so does Kingsolver acknowledge that people are much more difficult to fix than cars. Taylor gains mechanical knowledge as she works for Mattie but also comes to realize that while technology may carry one to the present or reveal the past, it can never erase that past. She ferries two refugees, fleeing a horrendous ordeal, to a safe home in Oklahoma, but she cannot restore their kidnapped daughter nor return them to their homeland. More significantly, she cannot erase the past of the Native American child she has acquired en route. The x-rays of Turtle, her adopted daughter, reveal the full extent of past abuse, and Taylor can't "stop thinking about the x-rays, and how Turtle's body was carrying around secret scars that would always be there" (127). Cars may provide limited sanctuary for the body but they cannot heal the body.

Road trips, of course, also expose the vulnerability of the body. The car offers some protection, but bodies on the road must be housed. Having just had a cold, wet child unloaded on her, Taylor realizes that her plan to sleep in her car must be abandoned, but she lacks the money for a hotel. So, she observes the various clerks in the various motels: "I drove by slowly and checked the place out, but the guy in the office didn't look too promising" (20). She hits pay dirt, predictably, with a "gray-haired woman" in the Broken Arrow Motor Lodge. Playing on compassion and appealing to standard female concerns—child welfare—she transforms the business into a home, convincing the woman to give her a room for free. While Sam revels in the comforting anonymity of a motel room that erases history and location, Taylor transforms motels from business establishments to domestic spaces. The Broken Arrow is far from an anonymous chain motel; Taylor sketches a vivid portrait of the family who runs it, familiarizing it and so reclaiming it as temporary domestic space, as family space. Thus, we see a twist on the hotel as "transit-place"; it becomes a domestic transit-place, opening up the possibility of conceptualizing domesticity itself as existing as a kind of transit-place. Rather than tying one to a particular space, domesticity can be found along the road, entered into and exited from (as Taylor leaves weeks later, "in hog heaven to be on the road again") at will (36). The culture of automobility creates the possibility for a kind of serial domestic space that does not ground or confine. Those with cars, an important qualification, can choose whether or not to participate.

Janet Wolff, building upon feminist concerns that just as women begin to come into subjectivity, theory begins to privilege a decentering of identity, has commented that, similarly, just as women begin to come into theory, theorists go on the road, privileging the metaphors of travel that she claims are inherently masculinist (234). But Kingsolver's work provides ways of thinking about mobility that may revise the masculine slant. She brings the car into the garden and introduces women who are not intimidated by cars. Taylor, moreover, is very consciously aware of her white American privilege. After talking to Estevan, forced to leave his country and his daughter behind, she realizes: "I thought I'd had a pretty hard life. But I keep finding out that life can be hard in ways I never knew about" (135). This is knowledge she now cannot retreat from, particularly once Mattie recasts it in automotive terms, insisting that Taylor drive a Lincoln when taking Estevan and his wife to a safer haven. Offended at the insult to her own car, Taylor is forced to listen as Mattie reminds her that the VW has a host of safety problems, and "just about any one of those things could get you pulled over by a cop. If you think you care so much about Esperanza and Estevan, you'd better start using that head of yours for something besides thinking up smart remarks" (184). One cannot negotiate a difficult political terrain without good car sense—or sufficient wealth.

Janis Stout, writing on women's narratives of departure, has identified the moment of departure as "an urge to move from a (relatively) private sphere of activity to a public and politically engaged one" (x). In departing from a life of legality to one of illegality—transporting undocumented refugees—Taylor drives into the public sphere at considerable risk to herself. Despite her realization that there are times when one needs to drive an expensive car, Taylor's resistance to the Lincoln also marks her resistance to standard American consumer culture. Although she enjoys the comfort and luxury the Lincoln provides, Taylor realizes that she misses her own car. In preferring her junky Beetle, she challenges the classic American impulse to attain bigger and better cars. While her ownership even of an old heap certainly privileges her above Estevan and Esperanza, she nonetheless forges a more nuanced road, suggesting the possibility of movement that is neither masculinist nor imperialist nor classist. Even Mason's Sam, who loves her own rickety Beetle, envies her mother's Trans Am, seeing it as the next step up the ladder of American consumer culture.

Both Sam and Taylor, however, recognize their cars not only as vehicles of mobility but also as domestic subjects. As David Gartman notes, early-twentieth-century automobile designer Paul Frankl sought to domesticate the machine, and so to eliminate "the fear people normally felt for machinery and its products so they will," in Frankl's words, "'no longer be restricted to the factory . . . [but] can now be admitted to Man's castle, his home'" (72). Women's ways of domesticating the machine function differently, transforming it not into the machine in the parlor but the machine as family. Naming cars, of course, is one of the most common forms of domesticating them, seen earlier by both Gertrude Stein and Rose Wilder Lane. Taylor names hers "Two Two" after the man who fixes the rocker arm. This positions the auto mechanic as potential kin, creating a kind of automotive family. The family connection is strengthened when Estevan and Esperanza take on the name of Two Two in posing as Turtle's birth parents and facilitating Taylor's adoption of her. Her adopted daughter is named after her car. While Sam never names her car, she clearly identifies it as an intimate, calling it "a good little bird" (6). Such close personal association with the car is totally lacking in *On the Road,* in which Dean Moriarty cares only about what feats he can make the car perform, not what he—or anyone else—*feels* about it. By turning the car into family (men may name cars but rarely figure them as family), these women blur the boundaries between human and machine, thereby opening a space for women's "modes of constituting knowledge and subjectivity," as Pratt has noted. In domesticating the automobile, they privilege women's expertise in a refigured domesticity. Once cars become domestic subjects, they no longer exist as wholly masculine vehicles.

This does not, however, transform them into vehicles of liberation. Taylor is well aware of the limitations of mobility and road trips. She may be "in hog heaven to be on the road again" (36), and Turtle "had spent so much of her life in a car, she probably felt more at home on the highway than anywhere else" (232), but she knows that cars—whatever their make and monetary value—do not offer any permanent sanctuary. Unlike Mason's Sam, who is constantly reminded of her exclusion from car culture, Taylor's story seems to function at times as a classic tale of female empowerment in a world more hospitable to women, even to women in cars. She just happens to run into motherly women hotel managers and supportive women mechanics. More than Mason, Kingsolver writes in the sentimental mode delineated by Jane Tompkins as "a political enterprise" (126), with her emphasis on the ways women's networks can enact social change. But Taylor is also forced to confront the limits of female empowerment. Turtle's short and unhappy life began, after all, in a car, both highlighting the car's function as domestic space and acknowledging the dangers of that space. The aunt who hands Turtle over to Taylor, apparently hoping to send her away from an abusive home, says, as proof that the child has no claim on anyone and can be freely handed over to a total stranger: "This baby's got no papers. . . . This baby was born in a Plymouth." She thus nullifies Taylor's earlier comment that "you got to have the papers and stuff. Even a car has papers, to prove you didn't steal it" (18). An unwanted and abused child holds a position of lesser status in the world of law and papers than even Taylor's beat-up old car.

The cultural valuation of cars over children renders it difficult to find solace and promise in a culture of automobility in which the car becomes home. As so many cultural and political commentators have observed, one needs a license to be able to drive a car but not to bring a child into the world. Car culture is thus associated with illicit sexuality and abuse. Even Taylor's admission of sexual experience draws on automotive culture. She may avoid pregnancy, but this "is not to say that I was unfamiliar with the back seat of a Chevrolet" (3). Pregnancy is not the worst that women have to fear from men in cars. Joyce Carol Oates's chilling 1970 story, "Where Are You Going, Where Have You Been?" presents a man in a gold car who seems to have complete knowledge of the details of his imminent victim's life as well as nearly satanic control over her behavior, luring her into the car to almost certain rape and death. Several more recent texts also reflect the very real dangers to women implicit in the equation of cars and sex; in Kathy Dobie's 2003 memoir, *The Only Girl in the Car,* and Beverly Donofrio's 1990 *Riding in Cars with Boys,* women's reputations are ruined and their lives are shattered by the results of both voluntary and forced sex in cars. Paula Vogel's 1997 play, *How I Learned to Drive,* details a history of incest masked as driving lessons. If metaphors of travel, as Wolff claims, are laden with male bias, metaphors of cars are weighed down with sexual innuendo, with potentially devastating repercussions.

Harry Crews, on the contrary, seems intrigued by the sexual potential of the car in his novel *Car.* When Herman eats a piece of the Ford Maverick, he "solemnly opened his mouth as though about to take upon his tongue a sacrament, but instead his pink lolling tongue lapped out of his mouth and touched metal, touched the hood of the Maverick car. It was clean and cold and he felt himself tighten around his stomach. He longed to have it in his mouth" (50–51). Yet even the hints of homosexuality do not imply danger. Rather, Herman feels that to have the car in his mouth "would amaze the world." For Crews the car is not so much a site for sexual activity (though it fulfills that function as well in the novel) as an object that conveys sexual gratification. Herman longs to make love to his car. Indeed, the whole novel is sexually charged. Mason and Kingsolver, however, in focusing on agency and mobility rather than sexuality, open up the possibility not just of cars as transportation rather than sex symbol but of women's lives shaped by agency and mobility rather than sexuality.

But all of this nevertheless constructs an automotive discourse that hardly seems liberating to women, who run the risk of pregnancy in cars, giving birth in cars, getting handed off to strangers in cars, and breaking down in cars. These cautionary episodes balance out Taylor's enthusiasm for what the car can offer and rein in her move toward empowerment, thus offering a model of a road trip that highlights its situatedness, one of the methods Wolff mentions for countering the masculine bias of travel theory. As Primeau notes, "American road narratives by women slow the pace, rechart the itineraries, and reassess the goals within the conventions of the typical road quest" (115). Lawrence goes farther, arguing that the difference in women's travel writing from men's "includes a strong sense of the constraints on self-propelled movement and mistrust of the quest and its purposeful destination" (20). Women on the road maintain a strong sense of positionality and limitations.

None of the women in these featured texts looks to escape from the home or from their family situations. Sam seeks greater understanding of her family background, and while Taylor seems more goal focused, she has no intentions of breaking her ties with her mother. Her initial trip may have been an attempt to escape domesticity and motherhood, but she accepts motherhood while renegotiating family and domesticity, finding it where the car takes her. Road trips are thus more integrated into women's lives. Jacqui Smyth has noted that "in mainstream American culture the quintessential fling before manhood—the fling that, in fact, transports one into manhood—is the road trip. The male tradition of the road is often used as a vehicle for discovering the self" (115). But women's road trips seek to determine the self's relation to the world and community and come to accept loss and even failure. Sam realizes she "will never really know" what happened in Vietnam, while Taylor voluntarily drives the man she loves to sanctuary with his wife. By recognizing their own marginality, whether from war or romantic love, they accept road trips as something less than transcendent while still gaining significant benefits from being behind the wheel.

Biker Chicks and Road Trips

Things change in Erika Lopez's *Flaming Iguanas: An Illustrated All-Girl Road Novel Thing*. This irreverent, and often raunchy, tale of a cross-country motorcycle ride overtly challenges the masculinist bias of the genre, first in its very form, as a "novel thing." Interspersed with illustrations, some with little bearing on the tale and many with very suggestive sexual innuendoes, the rambling tale bears scant resemblance to the linear form of a travel narrative. The journey originates by accident when the narrator, Tomato Rodriguez, accidentally runs over a cat and starts up a guilty friendship with the owner, Magdalena. Tomato impulsively suggests a cross-country road trip to be undertaken by the two of them (though they quickly part ways and she continues alone) in a newly formed motorcycle gang, the Flaming Iguanas. That she neither owns nor can drive a motorcycle does not deter her. Although she later justifies the trip by deciding to visit her father in San Francisco, the idea springs out of nowhere, unlike the more defined journeys undertaken by Sam and Taylor. But Tomato is much more aware of the symbolic implications of her quest as a means of asserting female agency and adventure: "Ever since I was a kid, I'd tried to live vicariously through the hocker-in-the-wind adventures of Kerouac, Hunter Thompson, and Henry Miller. But I could never finish any of the books. Maybe because I just couldn't identify with the fact that they were guys who had women around to make coffee and wash the skid marks out of their shorts while they complained, called themselves angry young men, and screwed each other with their existential penises" (27). By overtly challenging—and mocking—both women's exclusion from the road and the literary celebration of masculinity, Lopez's novel takes on and transforms a literary tradition of men on the road.

The most obvious difference between Tomato's trip and those of Sam and Taylor is that Tomato travels by motorcycle, a far more dangerous and culturally suspect form of transportation than the automobile. Women on motorcycles arouse much more suspicion and encounter considerably more resistance than women in cars. Barbara Joans and Michelle Holbrook Pierson, both authors of books on the motorcycle and cyclists themselves, note the prevailing assumption that women who drive motorcycles are lesbian, regardless of Joans's claim that women bikers are typically heterosexual—and white (88). Tomato, then, may partially fit the stereotype, as she comes out as a lesbian at the close of her journey, but this relegates her to the margins of "real" women's biker culture, which is largely heterosexual. And any woman, regardless of sexual orientation, does not find easy acceptance as a "biker," a highly privileged term. Joans cites a 1992 letter to *FogHog News* in which a man insists that women must "earn" the title by being able not just to ride but to "take a few boxes of Harley parts and build a strong running motorcycle, get your ass kicked a few times, end up in jails and hospitals any number of times, develop an asshole attitude" (qtd. in Joans 66).

With such hostility from both within and without the biker community, what makes women press for membership? Pierson asserts, "You may have to take my word for the fact that traveling by bike is superior to traveling by car" (151). She echoes the 1970s cult classic, Robert M. Pirsig's *Zen and the Art of Motorcycle Maintenance:*

> You see things vacationing on a motorcycle in a way that is completely different from any other. In a car you're always in a compartment, and because you're used to it you don't realize that through that car window everything you see is just more TV. You're a passive observer and it is all moving by you boringly in a frame.
>
> On a cycle the frame is gone. You're completely in contact with it all. You're *in* the scene, not just watching it anymore, and the sense of presence is overwhelming.   (12)

The motorcycle provides unmediated mobility. As an unenclosed space, it defies any kind of domestic association. Tomato, in many ways, drives a very different road, one even less conducive to women's traditional roles.

Flaming Iguanas transforms the female road narratives of Mason and Kingsolver, as it pays scant attention to reconfiguring domesticity or family on the road. Yet it does share an excitement regarding movement, even as it reflects Pritchard's caution regarding the progressive nature of mobility narratives. Asking herself if she feels like a "different woman" upon completing the journey, she replies: "No. I felt the same way right after I lost my virginity. / I figured and hoped the insights and profound sense of accomplishment would hit me later. . . . However, I must admit, I'm still waiting for the Womanhood Knowledge-stuff that was supposed to happen after I lost my virginity" (238). In equating the loss of virginity with mobility, Lopez aligns the sexual with the mobile and, tellingly, debunks each as a life-transforming event. Road trips do not remake your identity any more than sex does. While this may challenge the idea of mobility as powerful, it also identifies it as part of most women's lives, like the loss of virginity. Lopez thus claims the road trip as part of the normative female experience.

In addition to establishing that losing one's virginity does not qualify as a transcendent experience, she also challenges heterosexuality as normative. Having spent the novel uncertain of her sexual orientation—she has only experienced sex with men but fantasized about women—at her trip's conclusion she has blissful sex with Hodie, her father's female business partner, and openly announces herself as lesbian. Yet just like the completion of her journey (and the loss of her virginity), the experience does not transform her into a significantly different person. She has earlier lamented the fact that her trip seems to have had little impact on who she is: "I never got what I was looking for or where I was looking to go. I wasn't a good blue-collar heterosexual in a trailer home. I wasn't a real Puerto Rican in the Bronx. I wasn't a good one-night-stand lesbian. I wasn't a good alcoholic artist, and I wasn't a real biker chick because I didn't want the tattoos" (241). This catalog of identities comes off as both narrow and stereotyped: blue-collar heterosexuals live in trailer homes; biker chicks have tattoos; and so on. But after her experience with Hodie, she learns to eschew labels. "To my relief, the next morning I didn't feel like a member of a lesbian gang. I didn't feel this urge to subscribe to lesbian magazines, wear flannel shirts, wave DOWN WITH THE PATRIARCHY signs in the air, or watch lesbian movies to see myself represented" (251). This, then, constitutes her transformation into a person who no longer needs to seek group identity and can begin to define herself, resisting the cultural stereotypes she had fallen into. Sexual orientation does not fix identity; it allows one to acknowledge other options. In this way Lopez, like Kingsolver, avoids the trap of the sentimental belief in the possibility of political transformation through other women. While Tomato is delighted to have found sexual fulfillment, she does not mistake it for political awareness or even a sense of coming home to herself, continuing to refer to herself alternately as a lesbian and a bisexual. She does, indeed, seem to be the same exuberant woman who began the trip.

Tomato appears to privilege sexuality over ethnicity in terms of her self-definition, but this text also challenges the road trip as a white experience. While she has difficulty claiming her Puerto Rican identity—she doesn't speak Spanish, for example—it nonetheless looms large in her sense of self. She remembers "when I was twelve and started hanging around Puerto Ricans because I wanted to be one" (147). This comment indicates her belief that Puerto Rican identity is more than ethnicity; it means being part of a community. Her sense of disjunction between her heritage and her life troubles her: "Other times I wish I was born speaking Spanish so I could sound like I look without curly-hair apologies" (29). For her ethnicity seems to be constructed rather than essential, and she feels she lacks whatever it takes—whether language or group acceptance—to claim it. Yet even as she reveals her uneasiness about being Puerto Rican and her acceptance, with qualifications, of lesbian sexuality, there is one identity she asserts—that of a biker: "I felt alive and alone in the best way. No one could intimidate me or give me shit because I had bug guts all over me and could keep a bike upright and pass a truck in the crosswinds with a war cry. I'd just been through traffic hell and now I was actually a biker" (185). Far more than either Sam or Taylor, Tomato experiences the literal hardships of the road, of *driving*, as a means of self-expression and thus privileges her position as driver above ethnicity, sexuality, and even gender.

Her trip builds considerably on the revisions to the classic road narrative initiated by Mason and Kingsolver. Yet despite the greater physical exertion and danger involved, Tomato does not claim a macho position. She speaks openly of her terror of trucks and cars that could easily kill her, exposure to the elements, the discomfort of riding on female anatomy, and her dislike of being spattered with bug guts. Unlike Sam Hughes, who idealizes traffic, Tomato has a very realistic concern for her own safety. Further challenging any kind of posturing is the simple fact that she can barely ride, having acquired a bike the day before leaving and receiving only a couple of hours of instruction. Terrified, she creeps along at about ten miles per hour for the first several days, and for the whole of the trip she, like Gertrude Stein, never learns to go in reverse. Yet her very courage—or foolhardiness—in undertaking the journey strongly mitigates what could be identified as feminine weakness. While paying careful attention to vehicle safety, she dismisses the concerns of female vulnerability: "There is this myth that if you're a woman traveling alone people will instantly want to kill you.

This is an example of where you shouldn't listen to anybody. So much of the way we live and the decisions we make in this world are based on fear" (111). Tomato's self-confidence, even greater than Taylor's, allows her to "forget I'm a girl" and do what she wants, though she does have some advice: "The louder you laugh and the farther apart you plant your feet, the more respect you'll get. Take up space because it's not a school dance" (112).

This insistence on taking up space grants women a place in U.S. culture. Refusing to disappear, Tomato, a Puerto Rican lesbian biker chick, claims legitimacy and demands respect. She also asserts women's right to space, whether domestic or mobile, private or public. Like Sam, she revels in hotel rooms, seeing them as the ideal place. "Being in a hotel or motel was perfect living as far as I could tell. / There were no consequences. If you left a towel on the floor, you didn't have to pick it up. You never had to make your bed, and the sheets never smelled like sweat. Always fresh and white. I liked the paper band that was like a toilet hymen, and I got to be the macho conquistador" (133). While Taylor turns hotel rooms into a kind of transit domesticity, Tomato celebrates anti-domesticity: being able to drop towels on the floor and leave the bed unmade. In fact, motel rooms become a site of gender-bending, as she deflowers the toilet. This trip does not take domesticity on the road; it undoes any notion of women's domestic responsibilities. Yet by referring to the freedom from domesticity the motel provides, Lopez also reinscribes it as a force. One may enjoy brief domestic-free sojourns, but the towels and unmade beds are waiting at home. And while the rupturing of the "toilet hymen" may seem too silly to figure as a significant refiguring of female identity and agency, the playfulness of this text evokes a female picaresque novel. The macho road warrior becomes a Puerto Rican woman covered with splattered bugs and mocking male sexuality.

In fact, the novel closes with Tomato, a cartoon artist, conceiving a new project to earn enough money to fix her bike, which has broken down on her arrival. Hodie, who runs a sex toy company, offers her a job redesigning dildos. As Tomato envisions "groundbreaking penises" (256), her imagination takes off in typical fashion, imagining art shows of fake penises: "And just in case you blinked when the whole thing started, there'd be a best-selling biography on the trials and tribulations of the fake penises, inspirational fake penis stories of phony tragedy, pretend revenge, make-believe love, and counterfeit impotence" (257). In conceptualizing this fake-penis biography, Lopez reinforces her challenge of literary presentation first posited in the reference to Kerouac at the start of the novel. Once one undoes the boundaries of genre, narrative can accommodate a feminist reshaping of U.S. culture, privileging play above order and mobility above domesticity. By denaturalizing male sexuality, she decenters male dominance, leaving space for biker chicks to assert feminist agency. Tomato, very much her own woman, insists upon a woman's sphere in which going on the road reinforces femininity rather than domesticity and in which even a woman without a motorcycle license can learn to drive cross-country. Returning to Sidonie Smith's question of what being on the road makes of a woman "who is at once a subject as home and a subject at home," it becomes clear that in this novel being on the road makes a woman undefined by home.

The texts by Mason, Kingsolver, and Lopez suggest that women writers have seized upon travel literature and road trips, crafting conventionally male experiences into feminist texts. It follows, then, that metaphors of travel, despite Wolff's persuasive claim, may not be irredeemably masculine. The automobile provides a space for women to hit the road in a room of their own and, ultimately, to leave home behind. Kristin Ross has identified the automobile in France in the 1950s as the "center" of a "new subjectivity (whose circumference, unlike that of domestic subjectivity, is nowhere and everywhere), and of a new conception of nation" (22). For America in the 1980s one might revise this description to note that automotive subjectivity *becomes* domestic subjectivity, domestic transit-place, with a circumference both "nowhere and everywhere." By ungrounding domesticity without erasing it, cars and motorcycles significantly transform women as domestic subject. Rather than isolating them, the car brings them to a wider community, highlighting the position of the driver as both privileged and dependent. Patricia S. Mann has observed that with "the social enfranchisement of women, the scope of women's agency changes dramatically." In particular, she argues, "the quality of their domestic agency is altered, as well" (134). For these women drivers the automobile is not a fortress; it is a place of being simultaneously "at home" and not at home, of recognizing one's privilege and the cost it entails. As Biddy Martin and Chandra Mohanty note, feminists must negotiate between "the search for a secure place from which to speak, within which to act, and the awareness of the price at which secure places are bought, the awareness of the exclusions, the denials, the blindnesses on which they are predicated" (206).

The car and motorcycle can serve as such sites, particularly when driven by women such as Kingsolver's Taylor or Lopez's Tomato. Certainly, car travel is restricted to those with the means; while not absolutely limited to First World women, it remains true that the road trip is generally a Western phenomenon and one with a particularly U.S. twist. Yet its centrality in American culture justifies a reexamination while also acknowledging that being on the road is a privilege for the relative few. By moving into the driver's seat, women may exercise agency but are always reminded of their vulnerability in a technological age. In men's road trip narratives, says Smyth, "the road transforms into the 'home' of man" (115). Women's road trips transform the car into a domestic space and domestic subject, thus domesticating the road trip and challenging the limits of domesticity and mobility. The road may be the home of man, but the car revises the entire meaning of home. It helps to level the playing field by providing the means for propulsion yet carries an inherent reminder that unfettered agency and movement are neither available nor desirable. The female road trip thus constructs female identity as both mobile and situated, exercising agency and recognizing boundaries. In refusing to romanticize women on the road, these women writers open up the space for women with cars to follow new paths that can reshape gender and domesticity without necessarily denying them.

CHAPTER FIVE

Mobile Homelessness

Cars and the Restructuring of Home

In 1954 MGM released what would become its highest-grossing comedy to date, *The Long, Long Trailer*. Undoubtedly, much of the success of the film can be linked to the star power of its two leads, Lucille Ball and Desi Arnaz, then at the height of their popularity, based on the phenomenal success of their TV show, *I Love Lucy*. *The Long, Long Trailer* portrays a young newlywed couple, Tracy and Nicholas Collini (with marked similarities to Lucy and Ricky), negotiating the foibles of driving a trailer they have just purchased to serve as their home. In setting up this alternative home, the movie poses some interesting questions regarding what constitutes home in the age of the automobile. The lighthearted presentation of the trials and tribulations of mobile homemaking, however, anticipated some of the more serious issues surrounding the viability of the home that would command public attention in the later part of the twentieth century. With the freestanding, site-built home becoming an increasingly endangered species, the car often superseded the house as a site of domestic identity, as automobility replaced stability.

Women's road trips unsettled whatever assumptions about woman's place that still lingered in late-twentieth-century American culture. By hitting the road, women challenged the romance of American mobility and the shape of American domesticity. But that domestic space had already been splintered by the presence of the automobile. It is one thing to leave home; one can retain a belief in a home space, even if it is located elsewhere. It is quite another thing, however, to set up home in a car. Women's road trips may refigure domesticity, but they do not erase the idea of a grounded, situated home. When the car replaces the home, domesticity is not just unsettled: it is undone. Yet Americans are so in love with—and dependent upon—cars that even car homes hold a certain attraction, or at least we try to persuade ourselves that's so. By romanticizing the idea of the mobile home, we sugarcoat what may more appropriately be termed "mobile homelessness." Given the very real problem of homelessness in the United States, however, the car begins to look more and more like a viable, and even livable, alternative, a situation made possible by the long-standing appeal of the mobile in American culture, even of a house on wheels. James Clifford notes, "Once traveling is foregrounded as a cultural practice then dwelling, too, needs to be reconceived—no longer simply the ground from which traveling departs and to which it returns" (115). I would reformulate his assertion slightly: once automobility is foregrounded as a cultural practice, then dwelling needs to be reconceived.

Replacing the house with the car has obvious and, at times, catastrophic effects on women's lives. Yet most of the fiction that explores the relationship between house and car tends to privilege cars as providing a more flexible space. The gradual chipping away at the home as sanctuary, particularly in women's fiction, likely stems from many causes, from economics to domestic violence to having a more mobile culture. Kristin Jacobson has recently traced the development of what she terms the "neo-domestic novel," characterized by instability, self-consciousness regarding physical space, and recognition of the exclusionary power of the standard family home.[1] The genre begins, she asserts, in the 1980s with such texts as Marilynne Robinson's *Housekeeping* and Sandra Cisneros's *House on Mango Street*. *Housekeeping* opens with the mother committing suicide by driving a car into the lake. Choosing death in a car to life in a house indicates how fragile the home has become, and the novel closes with Ruthie and her aunt Sylvie trying to burn their house down. The young female narrator of *Mango Street* notes the discrepancy between the image of a large white house with stairs "like the houses on T.V." and what her family can afford. "But the house on Mango Street is not the way they told it at all" (4). The house may disappoint her, but she is thrilled with her developing female body, her hips "ready and waiting like a new Buick with the keys in the ignition" (49). Unable to find solace in a decrepit house, she turns to her body, indicating its potential by linking it to a car. The car functions as a metaphor of promise. A growing discontent with the traditional female sphere thus finds expression in the literature and is reflected in a realization that cars may provide something that houses do not—for a wide range of reasons. We must consider contemporary women's fiction against the backdrop of this reconfiguration of domestic space, now situated in both house and car.

The home has long been a source of power and a contested site in women's lives and women's literature. Much feminist analysis has focused on the so-called doctrine of separate spheres, either to explore how the home shapes female identity or to challenge the idea that such separation ever really existed, particularly for nonwhite and working-class women. These debates largely center on nineteenth-century American culture, in a time when the interior design of houses, according to Daphne Spain, contributed to gender segregation, given the divisions between smoking rooms and sewing rooms or libraries and parlors. In the twentieth century social conditions and a different style of home architecture ate away at those boundaries: "Whereas Victorian housing reflected a concern that each function and member of the household have a designated place, housing designs of the late twentieth century reflect the concern that no function or family member be limited to a particular space" (Spain 132). This reconfiguring left the car as an individual's sanctuary, a place for privacy that could replace the vanishing female space within the house. Joanne Mattera, in a 1990 article in *Glamour*, writes that "women are seeing their cars as a home away from home. Or at least a portable extra room" (255). As Lesley Hazleton notes: "They used to say that a man's home was his castle; it may be that a woman's car is hers. For many women, the car is where she is her own person, the place that is hers and hers alone. Where she can just be herself" (*Everything* 27). Even more than the house, which is likely to be in the husband's name, the car began to function as women's space.

The advent of the automobile reshaped the home in many ways, opening up the suburbs and allowing a greater distance between home and work. It also put the home on wheels. As auto camping gained popularity in the 1920s, trailers were devised to haul supplies, gradually developing into full-fledged accommodations. Both these developments—camping and trailers—reflect a response to concerns raised by women. Vacations, particularly family vacations, could prove difficult to arrange. If a family chose to travel, auto camping offered the best option, given that most hotels were not geared to accommodate women and children, having been largely devised as railroad hotels for commercial travelers.[2] But camping put tremendous demands on women to provide meals using rather primitive equipment and requiring a great deal of loading and unloading of supplies. Thus, trailers became a desirable improvement, cutting down on the loading time and often furnishing stoves to facilitate cooking. Because women were expected to make home on the road, trailers alleviated some of the domestic drudgery.

Originally conceived as vacation accommodations, trailers and cars became literal homes during the Great Depression, often to the consternation of those running campgrounds, as migrant workers and itinerants moved in on a semipermanent basis. Warren Belasco comments that the "highly visible automobility of these destitute people" was disturbing to locals and tourists: "At a time when cars were still considered signs of success, the migrant motorist was an unwelcome reminder of economic dislocation" (111–12). As cars became a necessity rather than a luxury, the sight of poor people in cars heralded the coming of a time when one could no longer count on houses, thus signaling a transformation of American culture. Using the car as home upset the normative assumptions about the role of the automobile, similar to middle-class indignation over the presence of luxury cars in poverty-stricken neighborhoods.[3] In providing both shelter and status, the automobile challenged the economic underpinning of consumer capitalism: that money represents worth. By approximating housing and status, the car offered lower-income people at least a semblance of middle-class comfort. Escalating housing costs put homes further out of reach, but cars were still within range. According to the U.S. Census Bureau, the median home price more than doubled from $47,200 in 1980 to $119,600 in 2000. While new car prices rose at about the same rate, to $25,800 in 2001 as reported by the National Association of Automobile Dealers, the same report indicates that the average price of a used car sold by a franchised dealer was $13,900. Given that used cars sold by private individuals tend to go for lower prices, it is clear that the car market accommodates a much wider range of incomes than the housing market.

Marjorie Garber's *Sex and Real Estate* explores the sexual attraction of the home yet confirms the realization that the house has become a luxury item because it focuses on the affluent, those who can actually afford either their dream homes or a reasonable facsimile of them. Richard Ford's 1995 novel *Independence Day,* however, counters some of Garber's optimism regarding the promise of the house. The protagonist, Frank Bascombe, who is a realtor, reveals "the one gnostic truth of real estate (a truth impossible to reveal without seeming dishonest and cynical): that people never find or buy the house they say they want. A market economy, so I've learned, is not even remotely premised on anybody getting what he wants. The premise is that you're presented with what you might've thought you didn't want, but what's available, whereupon you give in and start finding ways to feel good about it and yourself" (41). In other words, houses don't offer any symbolic satisfaction unless people learn how to make them meaningful, a task that becomes increasingly difficult in late-twentieth-century America. Thus, while the house may be "a primary object of affection and desire," a growing number of people must satisfy that urge with a car (Garber 207).

But low cost is only part of the allure of a car home. The romance of automobility also plays a role. In *The Long, Long Trailer* the husband is an engineer who travels to various jobs; his wife, Tracy, pushes him to buy the trailer as an alternative to sitting at home awaiting his return. She plans to travel with him and to set up home on the road, and the film is replete with images of the natural beauty available to them on the road, enhancing the notion of this honeymoon trip as a romantic adventure. Throughout the movie she refers to the trailer as home, overdecorating it to the point of danger, given the mountain roads they drive. Yet as much as she constructs it as home, its very mobility mocks her efforts. In a typical Lucille Ball slapstick scene she tries to prepare dinner in the moving trailer with predictably catastrophic results. She insists on collecting large rocks as souvenirs of their travels, nearly causing the overloaded vehicle to crash off a mountain. Clearly, she must learn to accommodate to this home on wheels, to acknowledge that while it may be home, it is not a grounded, situated place. The trailer, which nearly destroys the marriage, ends up as the site of reconciliation, as the couple agrees that their feelings for each other will allow them to adapt to a mobile home. Love, then, creates home in a typically comic—and conservative—conclusion.

Despite the happy ending, the issue of a trailer as home lingers, leaving viewers unconvinced of its potential to replace the house. The movie, in fact, attempts to re-write the history of the trailer as home. Although its origins were largely in the Depression, the growth of trailer parks and mobile homes exploded during World War II. As workers flocked to the factories, serious housing shortages were partly mit-igated by trailers, although John Hart, Michelle Rhodes, and John Morgan note that life in such trailer parks "could be a harrowing experience" (12).[4] The movie presents the trailer as a better and happier alternative to the conditions endured by migrant workers of the Depression and the war, capitalizing on the postwar return to nor-malcy sentiment that privileged home and women's return to it. Tracy's insistence on making home marks her as very much a part of that movement, though her driving ability sets her apart in a fascinating manner. The few times she is permitted to drive, hauling the trailer behind, she proves vastly superior to Nick, more skilled and less fearful. If women are determined to put the home on the road, they must learn to ne-gotiate the car as well as mobile homemaking skills.

Yet the picture of a luxurious—and expensive—trailer as a romantic home simply did not hold up as the century progressed. Largely associated with low-income, vio-lence, and low education, trailer homes continue to be perceived by the general pub-lic as undesirable housing, with the possible exception of upscale parks in the South for retirees. Their vulnerability to bad weather, lack of privacy, and precarious finan-cial and physical grounding seem to deny everything that a home is supposed to pro-vide. Yet in 2000 "mobile homes accounted for about 20 percent of all new single-family housing starts and about 30 percent of all new single-family homes sold" (Hart et al. 1). According to the U.S. Census Bureau, mobile homes constituted 7.6 percent of all single-family dwellings in 2000, up from 0.7 percent in 1950. The growing pres-ence of mobile homes indicates not just a housing crisis but a situation in which housing is increasingly destabilized—and mobilized. As Margaret Drury, in an early study of mobile homes, points out, American culture is caught between the appeal of the grounded and the mobile: "Our demand for permanence and our dependence on the land reflects our feeling of insecurity at being so mobile" (86). Thus, the mobile home may be a more "American" solution than we care to admit.

Mobile homes are not the only marker of the move away from the house. Such accommodations are luxurious compared to homeless shelters or the streets—or the car. While trailers may serve as temporary shelters in the wake of natural disasters, they have rarely been touted as a permanent solution to housing woes. Yet even that resistance is gradually breaking down. Homeless advocates in Austin, Texas, are beginning to experiment with housing people in RVs. A story about this effort in the *New York Times* opens: "On the eastern fringe of town, beside the airport and clogged freeway entrance ramps, there sits a bargain-rack version of the American Dream." Providing recreational vehicles for the homeless represents a hope that "owning a home, even one that is 8 feet by 32 feet, will anchor men and, eventually, women who have spent decades drifting through cycles of homelessness, substance abuse and failed relationships" (Healy). The notion, however, of the American dream as being defined by an "anchor" overlooks the mobile nature of American society and its growing homelessness. The number of homeless people in the United States is unclear, with various studies coming to various conclusions. But regardless of the actual numbers, it looms large in the public imagination; we are aware of the homeless to an extent unprecedented since the Great Depression. Also clear is that women and children are making up an increasing proportion of the total, though men still constitute the majority of the homeless population.[5] Yet there remains an aura—however misinformed—of the romance of the hobo, the man on the road because he wants to be, freed from all constraints. One thinks of Huck Finn lighting out for the territories or Sal Paradise and Dean Moriarty rebelling against bourgeois stability. In a 1996 article in the *Atlantic Monthly* Ian Frazier describes the time spent living in his van, a lifestyle he chose and remembers nostalgically. But he makes it clear that it is a male experience. In talking to car salesmen about which cars are best to sleep in, he notes: "All the sellers were male and over twenty-one years old. In my experience, anyone in that category out west has spent a certain amount of time living in a car" (48).

Living in a car may constitute a male rite of passage, but such a home remains unwomanly. Betty Russell cites studies indicating "that many men have enjoyed the freedom and lack of responsibility that homelessness can produce in them. The lure of the open road, the need to survive only for the day, and the opportunity to ignore rules have appealed to some men. . . . This is not the case with homeless women. Not one woman I interviewed and not one questionnaire reflected anything but despair and distaste for the situation in which the woman found herself. No woman wanted to be homeless" (82).[6] While Ruthie and Sylvie choose a life of transience at the end of Robinson's *Housekeeping,* few students, I have discovered, perceive their going on the road in the same light as they view the experiences of male itinerants. A growing awareness of homelessness makes it difficult to romanticize being on the road, especially for women, a lesson reinforced by the 1991 film *Thelma and Louise.* As Marilyn Chandler puts it, the "focus on house-building or home ownership as a completion of the rites of passage into maturity—and, more recently, on the mortgage as a token of stability—is still with us despite the increasing percentage of the population that cannot afford housing, let alone houses" (15). The house remains the standard by which Americans, particularly American women, are judged. No wonder cars begin to look more attractive.

Cars and Homeless Bodies

Given the increased visibility both of mobile homes and homelessness, the house seems a fragile site, out of reach for many and, for many others, not to be depended upon. Rosemary Marangoly George has suggested, in *The Politics of Home*, that home "is also the imagined location that can be more readily fixed in a mental landscape than in actual geography" (11). This may be true; a literal space, however, remains a necessity for literal bodies. Hence, the concern over finding a home—in a house, in the car, or on the street—remains paramount. In Marge Piercy's 1994 novel *The Longings of Women* Leila notes that "meeting former housewives, teachers, factory workers, waitresses, secretaries," makes her aware of "how fragile were the underpinnings of security for women. A man left or died, a job ended, a factory closed, a fire burned out her building, and she was out of money and on the streets" (428). If more and more women are being put out of the house, what impact does that have on these women's lives and in the literature that has a tradition of representing domesticity? What role does the car play? The Lynds discovered that "Middletown" residents in the 1920s were sacrificing money that could have gone into houses in order to buy cars, and that trend continues. Given the escalating cost of housing, many Americans choose to invest in cars instead.

Piercy's *Longings of Women* follows the intertwined stories of three women, one of whom, Mary, is homeless. Due to divorce, a tight labor market, and the gentrification that eliminates low-cost housing, she ends up on the streets in her sixties, working full-time cleaning houses but unable to save the money needed up front for an apartment. Yet her dreams center not on having a house but on having a car: "It must be wonderful to have a car, your home at your command. Just park it. It would get cold, and the insurance and gas and upkeep would be hard. But that was her fantasy, the goal of her savings. Then she would be like a snail, with her house always with her. She would never sleep in a garage or under a bush again" (42). Mary repeats this desire for a car at several points during the novel. Despite her homeless state, her dreams focus on a car, not a house. In her eagerness to be housed—or, rather, carred—she overlooks the potential perils of such a dwelling, a peril made painfully clear in William Kennedy's *Ironweed*, set in Depression-era Albany, where Helen must reluctantly accept the shelter of a car in order to survive. Her initial refusal to do so is clearly explained by what happens in the vehicle: "In Finny's car Helen would probably be pulling off Finny, or taking him in her mouth. Finny would be unequal to intercourse, and Helen would be too fat for a toss in the front seat. Helen would be equal to any such task. He knew, though she had never told him, that she once had to fuck two strangers to be able to sleep in peace. Francis accepted this cuckoldry as readily as he accepted the onus of pulling the blanket off Clara and penetrating whatever dimensions of reek necessary to gain access to a bed. Fornication was standard survival currency everywhere, was it not?" (89). The car may offer a modicum of shelter; it does not offer safety or any sense of home. Tellingly, Kennedy's focus is on Francis, Helen's partner, who must accept being cuckolded, rather than on Helen's fate in being forced to provide sexual gratification as "survival currency." A novel written primarily through Helen's perspective would describe the scene very differently.

Mary, though she fears rape on the street, seems to believe that the car will protect her, will provide walls to shield her from violation and violence. Her desire to be "like a snail, with her house always with her," evokes the image of a nomad, continually setting up home on the move. Rosi Braidotti, in *Nomadic Subjects,* sees such an existence as empowering, a postmodern condition: "Nomadism is an invitation to dis-identify ourselves from the sedentary phallogocentric monologism of philosophical thinking and to start cultivating the art of disloyalty to civilization" (30). Mary, however, does not have this luxury. As George reminds us, "The call for 'nomadism' and other such counter-travel brings to the surface the very structures of both imperialism and/or tourism that are so deeply entrenched in western gestures of leaving home" (36). "Leaving home" as an intellectual enterprise and being a nomad are two very different conditions. Playing with nomadism misrepresents the conditions under which "real" nomads may live, transforming it into a privilege for First World tourists— whether their touring is physical or theoretical.

Piercy, however, is not concerned in this novel with imperialism and travel; she explores the issues of homelessness at its most basic level: the woman who literally does not have a home. Mary may seek a kind of nomadism, particularly because, according to Braidotti, "nomadism consists not so much in being homeless, as in being capable of recreating your home everywhere" (16). But Piercy acknowledges, as Braidotti does not, that the ability to re-create home is more than an intellectual challenge. For Mary sleeping in church basements, abandoned buildings, airports, and, when she's lucky, the houses of clients who are out of town, the ability to re-create home is predicated on owning a car. This does not make her a nomad; it does, however, reflect the extent to which home has been refigured as portable property. The concept echoes Charles Dickens's 1861 novel *Great Expectations,* in which Wemmick, the lawyer's clerk, exerts considerable effort in the collection of what he terms "portable property," whether it be in the form of cash, jewelry, or foodstuffs. In these days of mobility one needs a portable home—and not only because it can move along with you. It is also considerably cheaper. The book is set in Boston in the 1990s, one of the highest-priced housing markets in the country.[7] Not only is a home financially out of reach for Mary; her occupation as a housekeeper further sours her on the desirability of having a house, which she views as too likely to disappear. Watching the marriage of one of her clients disintegrate, Mary assumes that the woman, Leila, will also lose her home: "Why did she call this Mrs. Landsman's house? Probably Mrs. L. thought it was hers, the way Mary had thought her house was hers, but it must be his, like the car and the money and the stock and everything else" (81). Houses cannot be counted on to protect women from the vicissitudes of fortune. They may be portable but primarily by men.

While Mary includes the car in the litany of what likely constitutes the husband's property, she nevertheless sees a car as within her reach. For a car is not just easier to afford; it also serves as female space in ways that houses used to. Chandler notes that the "'American dream' still expresses itself in the hope of owning a freestanding single-family dwelling, which to many remains the most significant measure of the cultural enfranchisement that comes with being an independent, self-sufficient (traditionally male) individual in full possession and control of home and family" (1–2). Indeed, Helen, in *Ironweed,* believes that if "Francis and Helen still had a house together, he would never leave her" (Kennedy 129). A house, she hopes, will solidify their relationship, indicating that she still measures success by the emblem of the home that is no longer available to her. But women such as Mary and Helen must revise the American dream that has betrayed them. Unfortunately for Helen, her own car is as out of reach as a house. Her death results as much from her unhoused exterior as from the cancer that gnaws at her interior.

Mary may be in a marginally better financial situation, with a few hundred dollars in the bank, but the only freestanding shelter that could potentially belong to her is a car, not a house, though that car offers both less and more security, as she remembers past driving experiences and how she was scared of trucks: "Trucks made her nervous, fearful they were going to drive right over her car. Enormous and male, they made her car feel small and feminine and in danger. She had a moment of longing for her last car, a green Datsun station wagon. It had been part of her" (141). Despite her sense of danger, the car was "part of her." Mary's description of her car's space—"small and feminine and in danger"—is an apt rendering of women's place, its vulnerability acknowledged yet also functioning as an extension of the self. The car certainly does not replicate the idyllic vision of the home; it does not lull Mary into complete security, as she remains fully aware of her fragility in a world dominated by "enormous" male power. It does, however, offer a brutally accurate picture of what it means to live in a culture in which one can be unhoused so easily. Cars keep you aware of where you stand; houses fool you into thinking that they are yours.

This novel illustrates the perils of continuing to rely on what has become, for many, an outmoded dream of the home. Becky, another character, murders her husband because divorce would mean she would lose the condo that constitutes her dream home. She relishes the times when she is home alone because it allows her the opportunity "to make love with the condo she adored" (241). She falls victim to the language of the real estate ads that Garber describes, of homes promising "not just a place but a relationship" (6). Becky, however, takes this relationship far beyond the light-hearted play that Garber envisions. The condo means more than sex to her; it means everything. Having grown up as one of seven children in an overcrowded house, the condo represents success to her, the home she's always wanted. As her marriage falls apart, she realizes: "If it wasn't for the condo, she'd just wave him out the door. Even the old car she drove was in his name" (301). Becky's unwillingness to give up on the American dream—as represented through a home—causes her to persuade her teen-age lover to help her commit murder. While she stands to lose the car as well, she acknowledges that her husband's family may permit her to keep it. But a car does not satisfy her. Unlike Mary, Becky refuses to rein in her desires, to consider the car as an alternative to the home. Mary, by the close of the book, owns a car and has a place to live; Becky is in prison, serving a life sentence. The inability to recognize that one's identity need not depend upon a fixed home causes her to lose her humanity—and, ultimately, her condo.

But replacing the house with the car is not a panacea. In *Longings of Women* it means accepting a lesser, more modest space because a home is too dangerous a dream. Longing for one can destroy a person. Piercy makes clear that houses can be deceptive; they can seduce someone into believing in their stability. And while Mary may believe that a car will solve her problems, the novel cautions that, at best, it provides only a fragile shell of protection. As even she acknowledges, the cost of a car does not end once one has purchased it. Richard C. Porter has investigated the various costs associated with driving, both social and financial, making it clear that owning and running a car can cost up to five thousand dollars a year (19). Mary's snail fantasy comes with a price tag she may not anticipate. There is no such thing as a free home. Moreover, her fantasy of car living is hardly an adequate answer for homelessness; "real" homeless women give different opinions of what it means to live in one's car. In trying to determine what constitutes being homeless—whether, for example, living with relatives or in a shelter counts—Jean Williams cites a woman who only considered herself homeless "when I was in my car. I didn't when I was staying in shelters because I had some place to go, a roof over my head" (143). Car tops do not necessarily qualify as roofs, though cars often provide the first line of defense against losing one's home. A 2004 Google search on cars and homelessness generated over a million and a half hits. Interestingly, when one inserts gender, a marked difference ensues. The phrase "living in her car" found 1,520 references, while "living in his car" produced only 989. Without putting too much stock in what is an admittedly unreliable source, it does indicate an interesting situation: even though men may be more likely to live in their cars by choice, and even though the number of homeless men exceeds that of women, women are apparently more likely to spend time living in their cars, possibly because even a car offers some shelter from the dangers of the streets.

Michelle Kennedy's 2005 memoir, *Without a Net: Middle Class and Homeless (with Kids) in America,* tells of her several-month sojourn living in her Subaru station wagon with three children, exploring not just what it means to be homeless but also what it means to live in a car. While she bemoans the discomfort of sleeping in the car—it "reeks of sand and beach and sweaty kids" (78)—she remains grateful both for its shelter and the power it confers to avoid spending money on cheap hotels, so she can save for the down payment needed on an apartment. Without the car she would remain homeless; with the car she finally manages to accomplish her goal. The car takes on multiple roles in shaping home. It becomes a literal home, the sign of her homelessness, and the way to a new home. As she says at the end, "I will never forget what it was like to drive home and be home all at the same time" (207). No house offers the same range of contradictory positions: safety and danger, mobility and stability, home and not-home. Women's lives have long since progressed beyond a simple equation of woman and house (if, indeed, such an equation ever really made sense); the car functions as a far more accurate symbol of what it means to be female in America, especially if one is homeless. As Linda McDowell points out, "Homelessness for women challenges every assumption about a woman's place and, for that reason, many women themselves often try to disguise or deny their predicament" (90). Michelle Kennedy uses a car to disguise her homelessness, worrying about what "kind of mother would put her children through this" yet knowing that within the culture of homelessness "what I am doing is for the best" (96, 73). The car is their best option in a difficult situation, providing the only home currently available.

One of the most important features the car offers, according to Kennedy, is freedom from homeless shelters. She vividly recalls a documentary on shelters "filled with drug addicts and gang members and families with poor little kids. The woman said that dirty diapers and needles were everywhere. God, at least we don't live in a city. At least I have the car. It might be awful to live in it, but it's still our own private space" (144). For Kennedy the car provides a kind of privacy and security unattainable in shelters, though the woman cited earlier only felt homeless in her car, not in a shelter. Clearly, there is no stable definition of what constitutes homelessness and what location provides the best alternative to it. The apparent numbers of women living in cars indicate that Kennedy is certainly not alone in being reduced to it. She even finds ways of making it acceptable. One of the car's advantages, she points out, is its ability to provide entertainment as well as shelter: "The kids and I drive all over the place. I love to explore, and it's a great way to kill time. After all we have a mobile home, why not be mobile?" (114). Valiantly doing her best to make the best of the situation, she learns how to take advantage of the car's precarious security. Like Lucille Ball in *The Long, Long Trailer*, she figures out how to make a mobile home.

Given the culture of homelessness in the United States over the course of the past twenty-five years, it is easy to see how "home" has become elusive, a contested site. With all the discourse—and controversy—surrounding women's getting out of the house, it becomes even more complex. As Spain has argued, the home is becoming devalued. The knowledge conveyed there has less status, and people spend less time in their houses—even as they build bigger and bigger dwellings (Spain 235–36). Less time in houses, more time in cars, particularly for women, caught between commuting to work and ferrying children to multiple activities. The number of female licensed drivers more than doubled between 1969 and 2001, and the number of vehicles has increased at an annual rate nearly one and a half times the number of licensed drivers (Hu & Reuscher 9, 10). The fact that the growth in cars exceeds the growth in drivers indicates how dependent we have become on the automobile; one seems to need multiple cars in an increasingly automotive age. The number of miles driven annually by men during that five-year period went up 49 percent, compared to an 89 percent increase by women (Hu & Reuscher 41). Even when not forced out of their homes, women are spending more time in their cars. Regardless of whether the car serves as a literal replacement for the house, it nevertheless functions as an alternative home, a space that both serves some of the same functions as the home yet also reshapes the meaning of *home.*

Reshaping the meaning of home can reshape identity. In Danzy Senna's 1998 novel *Caucasia,* for example, the loss of home encompasses loss of race and family. Birdie, the light-skinned daughter of an African-American father and white mother, is forced to accompany her mother underground, leaving behind her father and darker-skinned sister, Cole. Her mother, running from the FBI after becoming increasingly drawn into potentially violent protest groups in the mid-1970s, heads off in the car, convinced that the evidence left behind in her home will send her to prison. Leaving home proves devastating for Birdie, passing as white and living as a stranger in "Caucasia": "A long time ago I disappeared. One day I was here, the next I was gone. It happened as quickly as all that. One day I was playing schoolgirl games with my sister and our friends in a Roxbury playground. The next I was a nobody, just a body without a name or a history, sitting beside my mother in the front seat of our car, moving forward on the highway, not stopping. (And when I stopped being nobody, I would become white—white as my skin, hair, bones allowed)" (1). The car not only replaces home; it drives the body to a different race.

Homeless women have cited the loss of identity as one of the more devastating effects of homelessness. Maxine Harris notes that "when one becomes homeless, one's previous identity seems not to count, not to matter" (7). Betty Russell reports that after staying at a shelter for a few days to get a sense of what it meant, she found herself "becoming deeply depressed. In three days, no one used my name and I began to understand how one's identity can be lost. I was simply a homeless person" (28). For Birdie losing her racial identity marks her as homeless. Yet while the car may re-race the body, it also proves to be the only place where Birdie can be black. Outside the car she must put on a performance, pretend to be Jewish in order to shield her mother. If the Feds are seeking a white woman with a black child, Birdie's passing will stymie potential searchers: "The two bodies that had made her stand out in a crowd—made her more than just another white woman—were gone; now it was just the two of us. My body was the key to our going incognito" (128). Outside of the car Birdie's body turns white.

As illustrated in Chapter Two, the ways that race can be shaped by the car have their roots in modernism. In Jessie Fauset's 1928 novel of passing, *Plum Bun*, the privilege of the white world is reflected in the car that Angela's white racist boyfriend drives; when she is with Roger, she has the comfort of his car, as opposed to the public transportation she uses when on her own. The car offers privacy, while streetcars expose her to the public view—and to being racially constructed as black. Before deciding to pass, she finds it awkward to ride mass transit with darker-skinned friends. She "hated the conjectures on the faces of passengers in the street cars" (74). The car provided the space to be white in 1928, though, as Jack Johnson illustrated, it was no longer restricted to whites. In Russell Banks's 1985 novel *Continental Drift*, however, that space has become increasingly endangered. As Bob DuBois and his wife drive through Florida, they are struck by the numbers of nonwhite people: "They see them working in gangs in the orange groves, riding in the backs of trucks, mowing lawns, striding along the highways and sidewalks, and though Bob and Elaine are safely removed from these people, protected from direct contact by their car and all their possessions and by each other, the people of color seem up close and inescapably real, as if they are suddenly banging on the windshield, yanking at the door handles, climbing over the roof and hood and shouting to one another, 'Yo, man! Come on an' check out the white folks!'" (55). Clearly, the car's ability to preserve whiteness is rapidly decreasing, though it still seems to provide a very flimsy shelter. In Senna's 1998 novel, however, it allows Birdie to preserve her blackness: "But mostly my Jewishness was like a performance we put on together for the public. Only in the privacy of our car, on those long drives up and down the eastern seaboard, was I allowed to ask her about our real past" (140). Her "real" past and "real" race are confined to the automobile and put aside once they enter a dwelling. Privacy, then, takes place not in the home but in the car.

The car often functions as a literal home in this text: "Some nights the van had served as our home, parked in the darkened lot of a Roy Rogers, our bodies wrapped together in an afghan, limbs twisted around the other's for warmth. On particularly cold mornings the van wouldn't start at all and we just called the space where it had stalled our home, as good as any other" (143). In calling a stalled van in a parking lot as "good" a home "as any other," Birdie emphasizes the complete loss of home. The van serves as home because home, without her sister and father, is now primarily a place to sleep. Yet she grows attached to this life, probably because it allows for the possibility of Cole's return. As long as she keeps moving, she remains black—and Cole's sister. When she stops, she becomes white. No wonder that "staying still for too long felt unnatural." When her mother finally decides to settle down in a New Hampshire town, Birdie resists, knowing that the town would have no place for Cole and her father: "Our own place. The first in so long. But I tried to imagine my father and sister here, trying to figure out where they would fit in. Cole would say it was boring, a hick town. My father would say he couldn't live around these rednecks. They wouldn't want him, and he certainly wouldn't want them" (155). Home means the presence of her father—and, more important, her sister—and thus acknowledging her blackness. And because she can only preserve her racial identity in the car, only the car can serve as a site for home.

In some ways, then, race becomes portable, an entity that can be moved, like Mary's vision of the home as portable property in *Longings of Women*. Home and race are both characterized as mobile and material. Birdie's remaining material connection to her father and sister consists of a box of "negrobilia" originally left her by her father. Containing such items as the head of a black Barbie doll and a James Brown tape, it functions as a reminder of the African-American culture from which she is exiled. In adding to the collection with a Jim Rice baseball card, some hair from the only African-American girl in her New Hampshire school, and some information about Brazil, the purported destination of Cole and her father, Birdie tries to retain her past in much the same way that homeless women bring memorabilia with them: "Men tend to travel light, carrying items they consider necessary for survival. By contrast, the women tend to value items that either provide identity, such as marriage licenses or photographs, or security, such as a pillow or a souvenir, all reminders of a safer and more settled time" (Russell 103). Birdie's negrobilia provides both identity and security, reminders of a time when she was black and had a sister. For her the traces of her racial identity exist in these material objects, portable property. Race, like the car, is mobile.

Thus, while she and her mother settle down for a seemingly idyllic life in a cottage in New Hampshire, she favors her friend Mona's mobile home to the "gingerbread cottage, promising sweets and familial comfort" (146). Birdie prefers "trailer life to the world of my own house" (227). A trailer erases the false promises of home offered by the "gingerbread cottage" and openly admits that home is fragile, unstable, and mobile. Yet the people in Mona's world are unashamedly racist, constantly hurling racial epithets at the only two (acknowledged) African-American students in the school. Neither cottage nor mobile home can offer Birdie a space in which her body can be accepted. When her mother finds a boyfriend, she is even more bereft, feeling betrayed by the only person in her current life who knows and has embraced her as an African American. After learning of her mother's new love interest, she tries to console herself by watching cars go by: "My eyes stung as I looked at the cars swishing by before me. Country cars. Good-old American cars. Brady Bunch cars. Big brown station wagons with fake wood paneling. Suburbans, Buicks, Tonka trucks grown to life-size" (183). All of these cars, emblems of American domesticity, only remind her that her father "never had any faith in American cars." American cars define a Brady Bunch–like domestic scene that does not accommodate interracial, activist families.

Even the cars fail to offer comfort except in the potential for further movement. "Watching the cars pass by now," Birdie says, "I felt an urge to stick my thumb out and let some stranger take me wherever they were going. A trucker would be the best. We could just keep driving forever" (183). Only by driving forever can Birdie envision a life for herself. Cars, Paul Gilroy points out, "politicize and moralize everyday life in unprecedented configurations" and appear "at the very core of America's complex negotiations with its own absurd racial codings" (89, 84). The car both preserves Birdie's blackness and denies her place in American culture. Unhoused and un-raced, she is indeed homeless in America, "the easiest place to get lost" (1). Modernist writers such as Faulkner suggested that both race and cars were entities that could be made; Senna takes the association between blackness and automobiles still further in her presentation of the car as a racialized space, a place where race is perpetuated. The novel closes with Birdie tracking down her sister in Berkeley and seeing a busload of children, "black and Mexican and Asian and white," and, finally, a "mixed girl." "For a second," she says, "I thought I was somewhere familiar and she was a girl I already knew" (413). She feels, briefly, at home. But this idyllic vision of ethnic and racial diversity cannot redeem Senna's relentless exploration of what it means when one's body is not home any more than the gingerbread cottage can create home. The car appears to be the only way to recast home so that it can accommodate the complexities of racial identity in late-twentieth-century America.

The body is similarly prominent in Leslie Feinberg's 1993 novel *Stone Butch Blues,* in which transgendered Jess struggles to find home in her body and in a dwelling. Running away from her family's home at sixteen, she spends the next several years sleeping on friends' couches, living as an extra in other people's houses. This means, as she comes to realize, "I had no privacy here, no space anywhere in the world where it was safe to grieve" (157). If home is synonymous with safety, it is no wonder that Jess has such difficulty establishing home, for there is no safety anywhere in her world. Growing up in Buffalo in the 1950s, where it was still illegal for two women to dance together and where women were required, by law, to wear at least three pieces of "women's clothing," she is subjected to appalling brutality from the police and the general public. Initially identified as a "he-she," she ultimately begins taking hormones and passing as a man, believing that this will finally allow her to be "at home in my body" (171), a body she has not felt comfortable with since the onset of puberty. Yet even medical intervention to render her body similar to her own perception of herself as between genders fails to establish a sense of home, as she acknowledges in a retrospective preface: "*Strange to be exiled from your own sex to borders that will never be home*" (11). There is no home in her female body and no home in exile from it.

Physical structures offer no greater stability and security than Jess's own body. She works tirelessly to set up home at various stages of her life, a process that involves assorted home decorating projects. She finds a brief period of stability with Theresa, doing her best to replicate the conventional family house: "We got real furniture. I mean it was Salvation Army, but it was real. Our names were printed inside a heart on the dish towel that hung on the refrigerator door handle. . . . And there were marigolds in amber glasses on the windowsill, daisies in a green cut-glass vase on the kitchen table, fresh mint and basil growing in a flower box on the porch" (123). But despite the Ozzie and Harriet description, this replication cannot hold up against the pressures of gender. Once Jess decides that she must take hormones and pass as a man to survive, Theresa finds herself unable to participate in a charade of heterosexuality: "I can't pass as a straight woman and be happy. I can't live as the scared couple in apartment 3G who can't trust people enough to have friends" (152). It is only by subtly mocking home that home can exist. The accoutrements—daisies and dish towels— work only so long as one's sexual orientation is not masked. In some ways this illustrates Luce Irigaray's notion of playing with mimesis, by "playful repetition" to "convert a form of subordination into an affirmation, and thus to begin to thwart it" (76). By exaggerating the presentation of home and heterosexuality, this butch-femme couple calls into question the construction of both. Given that the home, according to Gill Valentine, tends to reinforce heteronormativity, with its master bedrooms and family rooms, it is only by queering the home, by exaggerating its place as a site of domesticity, that it can serve as a space of lesbian identity.[8] Once that queerness becomes too real, once Jess identifies as a man, that space is no longer viable.

The only place Jess finds safety, and thus home, is on her motorcycle: "Now this was the place I found my mobility and my safety—on this bike, under this helmet" (155). Like Tomato Rodriguez of *Flaming Iguanas,* she seeks a vehicle particularly associated with masculinity, the motorcycle. Unlike Tomato, however, she never identifies herself as a biker chick, though they both see how the bike functions as an emblem of sexuality. Tomato discovers "how much girls found riding on the back of my bike sexy. It wasn't me. It was the bike, the ride" (181). Jess remembers her first serious girlfriend, Milli: "I think I fell in love with her the moment she swung her leg over the bike and settled in behind me. The way two women relate on a motorcycle is part of their sex together—and she was very, very good on a bike" (106). In both cases the women claim the motorcycle not just as female space but as a site of lesbian sexuality.[9] Valentine notes that despite assumptions of sexuality belonging "in the private space of the home, not the public sphere," heterosexuality is constantly displayed in public, "institutionalised in marriage and in the law, . . . and is celebrated in public rituals such as weddings and christenings. This therefore highlights the error of drawing a simple polar distinction between public and private activities, for heterosexuality is clearly the dominant sexuality in most everyday environments, not just private spaces" (285–86). By bringing lesbian sexuality out of the closet and out of the house, these novels indicate that the vehicle continues to blur the boundaries between public and private. Displaying her sexuality on a motorcycle, without even the fragile protection of an enclosed car, Jess defiantly declares her bike a home space, the space in which one has intimate private relations. This goes beyond the destabilization of the home that Jacobson postulates as an element of the neo-domestic novel; here the home is in full view of the public, open to display.

It is also mobile. As Jess notes earlier, the bike is where she finds safety and mobility. In connecting safety to mobility, she makes an important point about how the concept of home has changed over the course of the century, particularly given the influence of automobility. Home has been mobilized, not just due to trailers and mobile homes but because the car makes it so easy to pick up and move that staying in one place becomes an outmoded concept. And given Jess's gender mobility—from girl to he-she to man and back to he-she—a mobile home is the only thing she can count on. While such a situation seems to fulfill what Braidotti sees as the promise of nomadism, "not taking any kind of identity as permanent" (33), for Jess it means not so much potential as vulnerability. A motorcycle, after all, offers even less protection than a car. Thus, it makes sense that when she is attacked for who she is, her bike rather than her home is the object of the attack. Seeing it destroyed by thugs who, unable to grab her, take out their rage on her bike, she realizes, "I knew it was only a motorcycle, but I felt like a ghost looking down at my own mutilated body on the asphalt" (157). The motorcycle, constructed as an emblem of butch identity in this text, thus functions as both home and body, as the house so often serves as an extension of the female body. A motorcycle, however, is much easier to destroy, reflecting how easy it is to become homeless. Jess, often shuffling among various apartments, in exile from her own body, is truly homeless, her only precarious space being on that motorcycle.

The motorcycle not only serves as home; it announces her identity to the world, as Jess declares: "All we got is the clothes we wear, the bikes we ride, and where we work, you know? You can ride a Honda and work in a bindery or you can ride a Harley and work at the steel plant" (100). Tellingly, she leaves home out of this litany of material self-definition. As a he-she, between genders, there can be no real home space. And as she will discover, the space of the steel mill is also closed to her; although the plant is forced to hire women, it is not forced to retain them. That leaves the clothes and the bike as markers of her identity. As she loses more and more—work, friends, and even gender after passing—Feinberg makes clear that the motorcycle remains as the only stable space in her life. Yet even in riding, she becomes an outlaw, unable to renew an expired driver's license because all her documentation marks her as female and she is now living as a man: "It was dangerous for me to ride—I'd been driving without a license for three years—but I lived to cruise on that bike. It was my joy and my freedom" (209). Jess is no longer defined by government documentation, that identification that can be so important to the homeless, proving one's existence in a hostile world. Rather, she looks only to her bike to assert her place in American society. But the combination of her hormone therapy and double mastectomy, the factories closing down, and the bar scene changing means there is no longer any space for her—or her bike. Consequently, when it is stolen, she finally decides to leave Buffalo for New York City, a decision she has been reluctant to make: "There was not much keeping me in Buffalo any longer. Yet I was still afraid to leave. I wanted to believe that whatever home I was looking for, I'd find it here. But the time had come to accept that my home might be waiting for me somewhere else. Or maybe I had to travel in order to find home inside myself" (225). With the bike gone, she realizes "it was time to leave Buffalo."

Jess is unhoused by the loss of her motorcycle, the one stable marker of her life in Buffalo's butch-femme community. A new life gradually opens up for her in New York, and the novel closes with her beginning to work as a union activist. But the text remains ambivalent about determining this new space as home. Indeed, while Jess revels in some of the opportunities she discovers in the city, such as music and books, we never hear her express about anything else the sheer joy she found in her motorcycle. Her new home, if anything, is even more dangerous and fragile than the old one, particularly without a vehicle to ground her. Certainly, the bike provided only a precarious home in a very specific time and place; in a very real sense there is no home on a motorcycle. But even its temporary sanctuary offered her something unavailable in any standard dwelling: a visible material representation of the transgendered self. Mobile homes more easily accommodate mobile identities.

Driving away from Home: Smiley and Simpson

Jane Smiley's novel *A Thousand Acres* (1991) has little to do with mobility. Rather, it focuses on what happens when one stays in the home and illustrates the extent to which the home fails in its time-honored role as woman's place. In this recasting of *King Lear*, Smiley explores the power of patriarchy and the ways that a family farm, a homestead, functions more as a prison than a sanctuary. Only by driving away from this grounded, situated, and oppressive space can Ginny Cook, the protagonist, find peace—and her own place.

Even on the farm, cars measure the value of the land. Ginny recalls the family's first car, noting that the neighboring children "continued to ride in the back of the farm pickup, but the Cook children kicked their toes against a front seat and stared out the back windows, nicely protected from the dust. The car was the exact measure of six hundred forty acres compared to three hundred or five hundred" (5). The farm's worth must be displayed by a car; the land itself is not enough to attest to the power that six hundred and forty (soon to be one thousand) acres conveys. Such grandeur demands a 1951 Buick sedan to reinforce its significance. Secure in the car, Ginny watches "the farms passing every minute, reduced from vastness to insignificance by our speed." Smiley sets up the perspective from the car as a potential escape from what appears to be the "vastness" of the farm. When trapped on the farm, however, the home space is anything but insignificant.

It seems easy to reduce the farm to insignificance because, given its spatial structure, everything is open to view. The mark of a good farm, according to Larry Cook, is "clean fields, neatly painted buildings, breakfast at six, no debts, no standing water" (45). The neatness, cleanliness, and standing water can be ascertained at a glance, and even debt is a known entity, as Ginny notes that "acreage and financing were facts as basic as name and gender in Zebulon County" (4). Everyone knows who has a mortgage and for how much. The same degree of neatness applies to the house itself. Farm women "are proud of the fact that they can keep the house looking as though the farm stays outside, that the curtains are white and sparkling and starched, that the carpet is clean and the windowsills dusted and the furniture in good shape, or at least neatly slipcovered (by the wife). Just as the farmers cast measuring glances at each other's buildings, judging states of repair and ages of paint jobs, their wives never fail to give the house a close inspection for dustballs, cobwebs, dirty windows" (120). Everything is visible on a farm; both interior and exterior are open to inspection. Although the rule of cleanliness connects inside and outside, Ginny makes clear that women's place is inside. The wife may work as hard as the husband, but her place is in the house. Indeed, she remembers her own mother instilling this notion into them as children: "We were given to know that the house belonged in every particular to her—that she was responsible for it, but also that damaging it was equal to damaging her" (223). As George points out, the "blurring of the distinctions between women, creativity and property is a trademark of patriarchal societies" (23). Women too, it seems, become portable—and patriarchal—property.

Such a close association between mother and home is not unusual in women's fiction. "The fantasy of home as mother," says Garber, "is a powerful and abiding one—the more so (as is often the case with fantasies) when the reality of the situation seems to fall short of the ideal" (48). And indeed, this house, so closely associated with the mother, is also the site of father-daughter incest. While the incest does not begin until after the mother's death, in some ways she facilitates its occurrence by insisting on absolute obedience to the father, by training her daughters to believe that their father is always right. The mother may see the house as her environment, but it also functions as the father's property—as do those who live in it. With the revelation of what lies behind the facade of the clean house, Ginny's earlier view from the car takes on a sinister hue. Speeding past houses and farms, one has no idea of what takes place inside; the clean fields may be on display, but the peril of the family home remains hidden. Regardless of the lack of visible dirt, there is much that does not bear close examination and certainly cannot be perceived from the car window. In fact, the new Buick that announces its owner's wealth and standing, in some ways protects him as a pillar of the community, a man beyond reproach. The car and the house work seamlessly to reinforce patriarchy, as even the female space of the home becomes a place of masculine power and abuse. Far from serving as an escape from the home, the car extends the oppression of the home. Even Larry's rash decision to divide his thousand acres among his three daughters is apparently sparked by a vehicle: envy of his neighbor's new tractor.

Harold's "brand-new, enclosed, air-conditioned International Harvester tractor with a tape cassette player" constitutes a challenge to Larry's superior acreage, and, even more galling, Harold refuses to divulge "how he'd financed the purchase, whether cold, out of savings and last year's profits (in which case, he was doing better than my father thought, and better than my father), or by going to the bank" (17). By withholding the financing information—a fact as basic as name or gender—Harold refutes the assumption that everyone knows everything about their neighbors and announces that there are things that remain hidden. This secret knowledge grants him privileged status, and the only way Larry can retake the advantage is to come up with a more elaborate financial plan: to form a corporation and distribute his land. A new Buick is no longer enough to reassert his supremacy, having been superseded by an even more imposing vehicle in the International Harvester.

Given the macho one-upmanship of this competition over both land and vehicles, it is not surprising that when Ginny drives away from the farm and the abuse it shelters, she drives her eight-year-old Chevy, a much more modest vehicle. When the car eventually breaks down, she trades it in for an eight-year-old Toyota with eighty thousand miles, appreciating "the way it looked in front of my apartment, unassuming and anonymous" (336). The car signals Ginny's resistance to the masculine power structure of the farm. In preferring an "anonymous" automobile, she undoes the ostentation of the new Buick or tractor. This car does not assert her place in the hierarchy; rather, it erases her, allows her to opt out. The fact that it is a foreign car reinforces its role as an alternative to the myth of the American family farm as idyllic space. Ginny finds her greatest safety in escaping the home and driving "unassuming"— and foreign—cars, not in claiming her share of the thousand acres. The place Ginny chooses is an apartment by the expressway where "all day and all night . . . you could hear the cars passing" (333). She is comforted by the traffic that is not seasonal or time-bound, a further remove from the farm. In the face of the horrors of the family home Smiley offers automobility—cheap cars and the sound of traffic—in exchange for the idealized American way of life on the family farm. Exploring the complex economic conditions that led to the loss of so many family farms in the 1980s, this novel reminds us that cars are what remain to offer us access to something else. An eight-year-old Chevy may sound like a poor exchange for a thousand acres of prime farmland, but the car hides no secrets and presents no false facades of normalcy. It seems to be all a woman can rely on.

The presentation of the car as an alternative to the home finds its fullest expression in Mona Simpson's novel *Anywhere but Here* (1986), in which Adele and Ann, mother and daughter, leave Bay City, Wisconsin, for Beverly Hills, seeking a film career for Ann and a rich husband for her mother.[10] Perennially short on money, they invest in the goods that present an impressive exterior: clothes and a car. But even before the high cost of living in California forces them to skimp on homes, the car trumps the house as the site for self-expression. They live for three years in Bay City in an unfurnished house where Ann is not allowed to open the refrigerator door with her hand because it will smear the chrome, and "sinks and faucets were supposed to be polished with a towel after every time you used them" (38). The house, in other words, is not for living. When the tension between her mother and stepfather rises in the house, she looks upon the car as a sanctuary: "Through the windows, the inside looked safe and closed and tended like a home" (71). The car seems so safe, in fact, that she ends up sleeping in it. The house thus functions as an unattainable ideal—at the end of the novel Adele claims to be seeking her perfect house, with a view of both ocean and mountains, far out of her price range—not as a space of domestic identity or safety. Indeed, domestic identity proves untenable in a world in which cars mean more than houses.

Complementing Garber's assertions about the love of houses, Adele falls "in love with a car" (114). Despite the phrasing, however, this love is not sexual. Unlike Tomato or Jess, she does not see the car as a sex object. She falls in love with what the car represents: status and the good life. When they sit in the car, a Lincoln Continental, the "leather moved below us, soft and rich. . . . It felt like our bodies would make permanent impressions" (123). One sees the sensuality of the car, the feel of the leather, but the significance lies more in Ann's belief that their bodies will make "permanent impressions," will mark the car as theirs, just as people mark their houses—marking that Adele does her best to erase in her preference for unfurnished houses and obsessive cleaning spells. The car, as it so often does, functions as an extension of the body, the skin meeting the leather to form a kind of literal bond. Thus, while Jacqui Smyth has identified the car as a "cover-up" for Adele, a sign of her escape from "domesticity while maintaining an acceptable exterior" (127), I would suggest that for Adele the car is more of a display, an announcement of who she is and what she wants. As she tells Ann: "We won't have the house, but we'll have a car to let people know who we are a little. . . . Maybe out there where everyone's in apartments, it goes a little more by the car" (122).

As the house loses its place—to trailers, to mobile homes, to shelters, and to apartments—other spaces take precedence. Smyth notes that in Simpson's novel "the car acts as a secondary, but more comfortable, home; it is a mobile dwelling that they park in front of the more luxurious houses in Beverly Hills in order to dream about the future home into which Adele hopes to marry" (127). Certainly, it serves as a mobile dwelling, but it constitutes a primary, not a secondary, home. While Ann tries to prevent her friends from knowing where she lives in Beverly Hills, she is proud of the Lincoln and always willing to claim it as hers. Her greatest admiration for her mother is expressed not in the series of dreary apartments they live in but in watching her mother drive: "Sometimes, I really liked my mother. She drove easily, with one hand, as she pumped the gas with the toe of her high-heeled shoe. We looped on the freeway ramps smoothly" (245). Adele can handle driving in Los Angeles; she proves less adept, however, at finding any kind of stationary home.

This is probably because, at heart, she doesn't really want one. The home represents a conventionality that Adele resists. As Dana Heller points out, "While houses represent the protective sphere of maternal stasis, cars suggest the public sphere of motion, the mobility of the father whose authority must eventually break the mother-daughter bond" (111). In this novel Adele is happy to relinquish the sphere of "maternal stasis" in order to claim the public sphere of motion. By inserting herself into the public, she recasts the place of the mother. If home is in the car and the car is in the public domain and the public domain is masculine, then Adele's loyalty to cars utterly undoes all the conventional ideology of domesticity. Her mobile life contributes, says Smyth, to "an unsettling of the domestic" (125). When the car replaces the house, domesticity does not just go on the road, as it does in women's road trips; domesticity is superseded by mobility, and the private gives way to the public, decimating any notion of separate spheres.

Adele's mastery of cars and driving marks her as a woman who will not be confined to the home, as a new kind of female hero, as Heller notes: "Tired of waiting for a hero, Adele becomes a hero. By stealing a Lincoln Continental, Adele wrests power from the masculine sphere and undermines its symbolic logic" (111). She may undermine the symbolic logic of the masculine sphere, but she also functions as a failed mother who puts her child continuously at risk due to her resistance to establishing any kind of stable home. We must be careful not to push her heroism too far; she is a dysfunctional human being who reveals the dangers for women of becoming too enamored of cars: one risks becoming endlessly mobile, ungrounded and unstable. The privacy of the house may enable incest for Ginny in *A Thousand Acres,* but the car keeps Adele in the public domain, unable to see beyond consumer culture as a means for self-definition. Simpson's presentation of Adele's choices of cars over all else invites both admiration and concern. When Ann sends her mother enough money finally to buy a house, Adele, predictably, buys a Mercedes station wagon instead, establishing her move to her next stage—being a grandmother: "And now I've got my station wagon ready for my grandchildren" (534). For Adele preparing for the various changes in her life means getting the right car. Although she claims to be waiting for that perfect dream house, it is clear that Adele prefers cars to houses. As Mary in *Longings of Women* realizes, cars are within reach; they're doable. Houses, on the other hand, have become a dream largely out of sight for an increasing number of American women.

Ann, however, has difficulty relinquishing her nostalgia for the family home and her desire for one: "Once, a long time ago, we had a home, too. It was a plain white house in the country, with a long driveway, dark hedges rising on either side of it. . . . It seemed then that the land around our house was more than owned, it was the particular place we were meant to be. Sometimes I thought we would stay there forever, that all the sounds of the yard would teach us about the world. But the trees never answered" (315). Although Ann is presented throughout the novel as a much more stable individual than Adele, Adele understands, in ways that Ann does not, that in today's mobile culture there is no place where one was "meant to be." Adele congratulates herself for her escape, crediting her "better" life with moving to California: "They tried—to make me and more than that, my child, into their mold. I had to let myself and my daughter go free" (531). Americans, of course, are particularly addicted to the notion of starting over, which often involves movement. "The automobile," says James Jasper, "is the perfect embodiment of this restlessness, the most seductive means of individual movement except for those archetypal dreams where you glide along without trying—better, perhaps, since cars are enclosed spaces, little homes you can take with you, where you can play music as loudly as you want, eat dinner, spend the night, even have sex" (4). The "little home" of the car, however, may forever keep one moving, unable to find refuge in a situated space.

Thus, while Ann returns over and over to her Wisconsin roots, she finally realizes, "I couldn't live there, I knew it" (488). The freedom conveyed by her mother may prevent her from regaining the sense of groundedness that characterized her early life, but it also prevents her from settling in with her aunt's anti-Vietnamese sentiments in Bay City and leads to a part in a TV show and an Ivy League education. Interestingly, Ann's mobility comes across as much more aimless than Adele's as she moves between Rhode Island and Wisconsin and appears to lack firm goals. Unlike Adele, Ann, while she certainly appreciates cars, does not invest them with the same significance. Thus, she is truly homeless, without a house or car. Although Adele has saved the Lincoln for her, Ann rejects it, consigning herself to a car-less life. This resistance helps her to break out of her mother's dependence on the material, but it also makes her a less compelling character. As problematic as cars may be, particularly in replacing the home, Simpson documents that they nonetheless provide women with a contested site for a complex, contradictory, and destabilized female identity, a very appropriate venue for establishing women's place in American automotive—and domestic—culture.

The presence of the automobile in American women's fiction reflects the crisis of home that has been raging in the last quarter of the twentieth century and into the twenty-first. The growing number of homeless women, a growing awareness of domestic violence, the decreasing ability to afford standard housing, and the increasing dependence on cars combine to illustrate that the promise of home is no longer valid. Increased technology—and increased surveillance—has rendered any notion of privacy within the home meaningless. Besieged by telemarketers, spammers, and radio and TV airwaves, the home is no longer a private space at the same time as the body seems to lose its fixity in a postmodern world. The advent of the automobile did not, of course, destroy the family home, but it does remind us of the dangers in relying on a situated place. That leaves the car. With little pretense of privacy, the automobile, nonetheless, may be the one place where many of us can be alone, the one place that may accommodate a body that has been defined as marginal by mainstream culture, whether because of race or sexual identity. The car's fragility and mobility proffers a more realistic location for home than the standard, grounded dwelling. The car, both private and public, gives the lie to the home as women's place. Women's place has become, for better or worse, the automobile.

CHAPTER SIX

Automotive Citizenship

Car as Origin

Cars define one's place in America; the car you drive reflects not just wealth or status but also something fundamental about your identity: new or used; American or foreign; sporty or utilitarian; SUV or hybrid. Even more than determining a sense of personal identity, however, the car functions as a marker of political and cultural belonging. As Daniel J. Boorstin has observed, "The automobile has been the great vehicle of American civilization in the twentieth century" (vii). Back in 1921, President Warren G. Harding remarked that "the motorcar has become an indispensable instrument in our political, social, and industrial life" (qtd. in Flink, *Car* 140). While it may have lost some of the allure it used to carry, as more people take the car for granted, it continues to serve as an "indispensable" icon of American identity. "It's the '49 Ford by a landslide," declared a Ford ad in a sly dig at the presidential election of 1948, nearly a dead heat between Truman and Dewey. The Ford embodies the triumph of the American political system more thoroughly than President Harry Truman's narrow victory. By the 1950s you could "see the USA in your Chevrolet," and in the 1960s we were reminded of the litany of what defines basic American values: "baseball, hotdogs, apple pie, and Chevrolet." Joyce Carol Oates, in her 1969 novel *them*, also illustrates the hold the automobile has on American identity through her protagonist, Jules: "So long as he owned his own car he could always be in control of his fate—he was fated to nothing. He was a true American. His car was like a shell he could maneuver around, at impressive speeds; he was second generation to no one. He was his own ancestors" (335). Jules overestimates the car's ability to allow him to control his fate, but his sense of himself as a "true American," his "own ancestors," reflects a very American belief that our originary American identity stems from the car.

This dependence on the car comes, of course, at the cost of other values. As Lewis Mumford lamented in 1963, "The current American way of life is founded not just on motor transportation but on the religion of the motor car" (234). Roland Barthes's famous comparison of the car to Gothic cathedrals acknowledges the same point: "I think that cars today are almost the exact equivalent of the great Gothic cathedrals: I mean the supreme creation of an era, conceived with passion by unknown artists, and consumed in image if not in usage by a whole population which appropriates them as a purely magical object" (88). Indeed, in Don DeLillo's 1985 novel *White Noise*, as Jack Gladney watches his sleeping daughter, she mutters, "Toyota Celica." "The utterance," marvels Jack, "was beautiful and mysterious, gold-shot with looming wonder. It was like the name of an ancient power in the sky, tablet-carved in cuneiform" (155). The higher power that guides contemporary culture derives as much from the automobile as from any supreme being.

Even in the early twenty-first century, with religion becoming an ever more significant force in American public life, cars retain their reverential status. Denise Roy, searching for the time and space for faith in an increasingly hectic and mobile life, declares, in *My Monastery Is a Minivan*, "A minivan might not be as good as a monastery for finding peace and quiet, but it is precisely the place where I find the face of God" (15). We may have lost some of our initial enthusiasm for the automobile, and most of us probably view it in a more secular light than Roy, but its time is far from over. Examining the "decline" of the auto-industrial age in the early 1970s, Emma Rothschild nevertheless admits that "automobiles still serve as a focus of national emotions" (74). Clearly, the death of the automobile, claimed by John Jerome in 1972, has been grossly exaggerated. As Jean Baudrillard observed in the 1980s, by driving in America, "you learn more about this society than all academia could ever tell you." American cars and drivers, he goes on, reflect American identity (54).

Cars do more than reflect American identity. Cars can determine a kind of American citizenship. Particularly for those perceived as outside of the mainstream based on race, class, gender, and ethnicity, the car can replace home and nation in shaping one's place in American culture. It constructs what I term "automotive citizenship," a condition in which one's sense of belonging rests largely on the car, both literally and symbolically. For the car, as the vehicle that spans the private and public realms—that can serve as an extension of the home—provides a flexible and unstable location from which to negotiate among the various sites that constitute American identity. The role of the automobile holds particular force for those who may lack more solid standing within the community: "white trash," immigrants, nonwhites, and, most powerfully, women within such subgroups, for it can serve as a bridge between dominant and nondominant cultures. One may not be able to drive into the mainstream, but the car facilitates movement among different locations, both physically and figuratively. So, while automotive citizenship does not construct an idyllic space, it does offer an alternative to those denied access to American civic culture. It situates an individual in a kind of borderland but one grounded in material reality; cars, after all, demand fuel and maintenance and rely upon the actions of real people.

Cars are both material objects and symbolic icons. And if we have learned anything from the past thirty years of feminist scholarship, it is that bodies matter, that women's lives are shaped by material circumstances. Automotive citizenship may constitute a hybrid identity, but that hybridity negotiates a site between the body and a poststructural theoretical self. As Caren Kaplan observes, "The self is always somewhere, always located in some sense in some place, and cannot be totally unhoused. New technologies appear to promise ever-increasing degrees of disembodiment or detachment, yet they are as embedded in material relations as any other practices" ("Transporting" 34). The same holds true for older—automotive—technologies; the automobile industry may rely on crash dummies for safety tests, but cars rely on human bodies.

That cars permeate American culture hardly needs proof. The automobile has reshaped the American psyche and the American landscape. As much as the clothes we wear or the people we associate with, the car announces our presence. We are what we drive, and driving confirms our American identity. Author Studs Terkel relates an incident while trying to cash a check at a hotel. He doesn't have a driver's license and offers his library card instead: "Then they knew I was un-American right there. A driver's license is your ID. That's your mark of identity" (qtd. in Coffey 296). Terkel's claim is prescient; in May 2005 the U.S. Congress passed a bill that, in effect, makes the driver's license into a national identity card. According to the *New York Times*, "A state would have to require proof of citizenship or legal presence, proof of an address and proof of a Social Security number. It would need to check the legal status of noncitizens against a national immigration database, to save copies of any documents shown and to store a digital image of the face of each applicant" (Wald & Kirkpatrick). The driver's license has become such a necessity in American culture that it has become a mark of citizenship and legality.

To be American is to drive; it is also to have a car. "An American without a car is a sick creature," observes Andrei Codrescu, "a snail that has lost its shell" (3). Despite concerns over pollution, traffic, dependence on foreign oil, and the paving over of the American landscape, Americans remain unwilling to curtail their automobile use. Twice in recent years America has gone to war at least in part out of concern that Middle East instability could disrupt access to valuable oil fields. In *Highways to Heaven: The AUTO Biography of America* Christopher Finch contends, "As is exceedingly clear from the history of the automobile in America—its rise to a position of dominance in the transportation field—the average citizen equates the car with personal freedom, which is held to be a self-evident and inalienable right guaranteed by the Constitution" (372). People take this constitutional "right" seriously. Sociologist David Gartman recalls, "My father's identity depended fundamentally on cars—'I'm a Chevy man,' he resolutely declared, and never bought anything else" (xiii). This type of dedication to the automobile indicates that the car goes beyond typifying American identity; the car shapes American citizenship. But both cars and citizenship tend to evoke a masculine tradition. It is, after all, the father in Gartman's quotation who is defined by his Chevy. If to be American is to claim a kind of automotive citizenship, does such a concept reinforce women's second-class status? Often subjected to violence within cars, often excluded from the driver's seat, women are as frequently victimized as served by the automobile.

The discourses of citizenship have long troubled feminist theorists, who contend, as Kathleen B. Jones puts it, that "political theorists have, for the most part, developed a conceptualization of citizenship in the West by representing the identity, actions, and locale of a dominant group of elite political actors—white, heterosexual, bourgeois, European men—as the model of citizenship for all" (2). Indeed, in 1938 Virginia Woolf implicitly acknowledged this masculinist privileging with her famous comment: "As a woman, I have no country. As a woman I want no country. As a woman my country is the whole world" (109). My interest, like Woolf's, is less in legal citizenship (though that enters into my later discussion of Viramontes) than in a kind of cultural citizenship, a feeling of national identity and belonging not defined by documentation or legality but in part determined by the car. Adrienne Rich responds to Woolf's comment by reminding us that we cannot "divest" ourselves of our country and cautioning that to refigure our sense of belonging, we need to "begin with the material" (212, 213). I contend that the car, a material object, opens up an extralegal space of citizenship, confers a sense of belonging, and grants access to public, civic life. May Joseph suggests, "The citizen and its vehicle, citizenship, are unstable sites that mutually interact to forge local, often changing (even transitory) notions of who the citizen is, and the kind of citizenship possible at a given historical-political moment" (3). The car is yet another unstable site that reshapes access to the larger community, that indicates who belongs and who doesn't. Particularly in novels by nonmainstream writers, novels in which citizenship is a highly contested sphere, the car often supplants home and even national origin as the site of identity and identification. While certainly not restricted to women, issues of automotive citizenship resonate particularly strongly for women; automotive citizenship refigures home, elides boundaries, revises standards of inclusion, and foregrounds the body, all of which tend to be critical concerns in women's fiction. These texts explore an automotive alternative to mainstream civic culture, opening up new ways of thinking about U.S. identity.

One of the key issues in rethinking citizenship lies in its public nature. The citizen is a public figure and, given the extent to which the public sphere continues to be dominated by white men, citizenship thus takes on a white masculine character. But that very gendering of public and private has been forcefully challenged by feminist scholars. Virginia Scharff's *Taking the Wheel*, for example, illustrates the way that the automobile itself undoes such a gendered separation: "For women, the journey from back to front seat, from passenger-hood to control of a vehicle deeply identified with men, began in public and entailed repeated confrontations with popular manners, morals, and expectations" (17). The car propels women into the public sphere and a public position, challenging popular assumptions about women's place. Ruth Lister examines "the rigid, gendered separation of the 'private' or domestic and 'public' spheres, in which the former has represented particularity, care, and dependence and the latter universalism, justice, and independence. The rearticulation of this public-private divide provides one of the keys to challenging women's exclusion at the level of both theory and praxis" (22). One way of rearticulating the public-private divide lies in the car's ability to challenge the division, serving as a vehicle that collapses such distinctions. The automobile, so crucial in granting women greater access to the public sphere early in the century, continues to elide the boundaries between private and public; as detailed in Chapters Four and Five, cars often become domesticated, sites of home, in women's literature. By destabilizing the public-private divide, cars offer a space for citizenship both private and public and are thus especially befitting women's borderline status. The automobile, then, provides a means to reconfigure notions of citizenship, belonging, and identity for women, opening up a space for automotive citizenship.

Allison: White Trash Automotive Citizenship

A sense of American identity does not come automatically to those born in the U.S.A., even if they're white. Dorothy Allison, exploring how class and sexual orientation may deny a person access to mainstream American citizenship, perceives that the car often defines "white trash" subjectivity in U.S. culture. One of the most common stereotypes associated with so-called white trash identity is the appearance of cars on the lawn. John Hartigan, analyzing the construction of white trash in the Detroit area, quotes community residents complaining about a "white trash family" who has moved into their neighborhood: "They've got four cars in their backyard, and he's always messing around with them. Then they got two parked out on the street that don't run" (46). Other interviewees make similar complaints: "We've got white trash moving into this neighborhood. . . . [T]hey park a vehicle right on the front lawn; they do car repair right in the garage" (49). Cars, particularly cars that seem to function as more than transportation, mark class status and even identity: cars on the lawn equals white trash identity. Bone, the protagonist of Allison's *Bastard Out of Carolina* (1992), reinforces and embraces this characterization as she lovingly describes her uncles who "tinkered with cars together on the weekends, standing around the yard sipping whiskey and talking dirty, kicking at the greasy remains of engines they never finished rebuilding" (22). Bone yearns for the houses her aunts and uncles live in, where "nobody cared if you parked your car on the grass" (80). But she finds herself imprisoned in houses where her middle-class stepfather yells at the uncles "if they drove up on the grass" (81). While cars serve as only one of many class markers in this novel, they clearly reinforce the family's white trash status.

And this status is precisely what challenges their citizenship, their belonging. As Allison tries to explain in her 1994 collection of essays and addresses, *Skin:* "Entitlement . . . is a matter of feeling like we rather than they. You think you have a right to things, a place in the world, and it is so intrinsically a part of you that you cannot imagine people like me, people who seem to live in your world, who don't have it" (14). Allison recognizes that legal citizenship does not necessarily confer acceptance and belonging. Allison's feeling that she has no right to "a place in the world" is illustrated in her novel partly through the subtle ways that cars help to shape Bone's uneasy status. As an "unstable site" of citizenship, cars fix her as marginal within the dominant middle-class culture, yet they also suggest a different way of framing identity, one that reconfigures what it means to hold citizenship. Automotive citizenship may offer an alternative to the civic society that defines her legally as a bastard.

Cars shape Bone's life in far more specific and personal ways than simply as class markers. Born prematurely as a result of a car accident—which her mother sleeps through—she is certified a bastard by the state of South Carolina. Even this highly unusual three-day sleep comes courtesy of Uncle Travis's Chevy, "which was jacked up so high that it easily cradled little kids or pregnant women" (1). Having gone through eight months of pregnancy with virtually no sleep, Anney succumbs to the comfort of a retooled car and sleeps through being flung through the windshield and even giving birth. Unable to speak up and bluff her way through the hospital's record keeping, she thus loses the opportunity to claim legitimacy for Bone. While the car clearly contributes to the official designation of Bone's illegitimacy, her grandmother seeks to embrace rather than challenge that status: "Granny said it didn't matter anyhow. Who cared what was written down? Did people read courthouse records? Did they ask to see your birth certificate before they sat themselves on your porch? Everybody who mattered knew, and she didn't give a rat's ass about anybody else" (3). By challenging legal definitions, in which she is supported by her son Earle—"The law never done us no good. Might as well get on without it" (5)—Granny opens up a space for a different kind of citizenship, one beyond legality.

Granny's implied alternative to courthouse records postulates a way of belonging that echoes what May Joseph calls "nomadic citizenship": "As both an imposed condition and process of negotiation, nomadic citizenship suggests the ambivalent, lucrative, unconscious, and itinerant ways in which migrant subjects live in relation to the state" (17). The members of the Boatwright family do not necessarily qualify as migrant subjects in terms of transnational migration, but they are always on the move, changing houses, changing jobs, moving children from one nuclear family to another. Further, as white trash subjects, they certainly have an itinerant relation to the state. With illegitimacy imposed upon Bone and the family's attempts to reconfigure what constitutes legitimacy—as Granny says, "An't no stamp on her [Bone] nobody can see"—Allison situates the body as a means of challenging the state (3). Bone's body, at least, has no stamp of illegitimacy. The discourses of citizenship thus negotiate among the state and the family, with the body, produced by a car wreck, finally reconceptualizing legitimacy. But the car does more than set this process in motion; by serving as a marker of white trash identity, it reminds us of the various ways identities are shaped and imposed. Given the car's power to confer status and place, it becomes aligned with the forces that determine belonging. In embracing her white trash status and maintaining an unstable relation to the state, Bone also embraces a world at least partly defined by the automobile.

Indeed, most of the significant moments in Bone's life, from birth onward, occur in cars. Her future stepfather, Daddy Glen, finally persuades Anney to accept his proposal while Anney is sitting in her car with her two daughters. The physical awkwardness of the situation foreshadows Glen's abusive behavior. He reaches "through the window to take Mama into a tight embrace" and then pulls "Reese and me forward . . . so that Reese's bird bones crunched into my shoulder." Still more tellingly, his face "pressed into mine, his mouth and teeth touched my cheek" (36). Given the violence implicit in the description, it should come as no surprise that the first time Glen abuses Bone, the incident takes place in the car. Most tragically, after Glen has brutally raped and beaten Bone, Anney pauses before driving her to the hospital to embrace Glen as he beats his head against the car door in a frenzied fear that he has finally gone too far and alienated Anney beyond recall. Observing the interchange from the passenger's seat, Bone comes to the most painful moment of her existence: "Rage burned in my belly and came up my throat. I'd said I could never hate her, but I hated her now for the way she held him, the way she stood there crying over him. Could she love me and still hold him like that?" (291). As so many women have discovered, cars provide no safe space from either sexual assault or emotional betrayal. Automotive citizenship reminds women of their continued vulnerability in a world dominated by male violence. Yet in so doing, it also reinforces the position of the female body; amid discourses of place and identity one cannot overlook the body. First and foremost, we live and are defined by our bodies, as Allison underscores by her graphic description of the violence done to a thirteen-year-old child. While Bone's body may not reveal her illegitimacy, it does reveal the scars of physical violence.

If this constitutes automotive citizenship, who needs it? I am not suggesting, however, that the car facilitates violence. Rather, the fact that so many defining—and painful—moments take place in the car reflects the ways that the car, rather than the house or the community, seems to be the space in which people are shaped and determined. Given the migrancy of the characters who move houses every few months, it makes sense that the car becomes the site of origin: the origins of illegitimacy, of abuse, of betrayal. Houses are temporary in this text; cars are always there. In fact, on first meeting Glen, Anney thinks: "I need a husband. . . . Yeah, and a car and a home and a hundred thousand dollars" (13). Significantly, the car trumps both home and money in terms of necessity, as it precedes them in her list. Unfortunately, a husband seems to top her priorities, helping to explain how she can remain with a man who abuses her daughter. Just as automotive citizenship insists upon the presence of the female body, it also reminds us of other types of liabilities for women when the perceived need for a husband and for romantic love can trump what many would identify as woman's "essential" nature: to protect her young. By denaturalizing maternity, Allison opens up a space to examine all the complexities of women's lives, refusing to accede to "types" of womanhood. Certainly, the automobile has added to the complexity—and completeness—of female identity in the twentieth century.

Bone's association with automotive citizenship both acknowledges and challenges the role of family in conferring place within the larger culture. Being a Boatwright defines her, and she closes the novel identifying herself as a "Boatwright woman" like her mother. But given the physical and emotional assaults committed against her in the family car, we also understand that family has its limitations. Citizenship has become closely aligned with family in recent political discourse. Lauren Berlant argues that in the 1980s, under President Ronald Reagan, citizenship shifted from a public stance to a private one in which intimate things such as sexuality and so-called family values are "key to debates about what 'America' stands for, and are deemed vital to defining how citizens should act" (1). Such a move idealizes both family and the private sphere, creating what Berlant terms a "new nostalgia-based fantasy notion of the

'American way of life'"; this "residential enclave where 'the family' lives usurps the modernist promise of the culturally vital, multiethnic city" (5). But Allison has established that those "residential enclaves" where family values are constructed (and deconstructed) may be cars rather than houses. Thus, automobility shapes the intimate concerns that constitute citizenship and so transforms the private sphere. For the car, of course, is the vehicle that gets people out of the house, the vehicle that was so instrumental in moving American women into the public sphere. This sets up a type of nomadic citizenship but with a twist: it becomes automotive citizenship, muddying the distinction between private and public and thus implicitly constructing a space for women's citizenship that partially depends on erasing those boundaries. Despite the horrific abuse that occurs in the car, then, the very fact of the car as significant space reformulates the possibility of a construction of subjectivity that does not relegate women to second-class citizenship, even if it leaves them physically vulnerable.

Automotive citizenship is certainly not idealized: witness what happens in cars. But it does offer the chance for new movement, for a new reconfiguration, a new negotiation of subjectivity and possibility. Significantly, Bone constructs fantasies of movement as a means of coping with her situation: "I began to imagine the highway that went north. No real road, this highway was shadowed by tall grass and ancient trees. . . . Cars passed at a roar but did not stop, and the north star shone above their headlights like a beacon. I walked that road alone, my legs swinging easily as I covered the miles. . . . Only the star guided me, and I was not sure where I would end" (259). Here Bone conflates the Underground Railroad with the highway, as she envisions a flight to freedom. In many ways this passage evokes a pre-automotive age with a road shadowed by ancient trees. Yet even this seemingly ancient road has cars, which seem to be guided by the North Star, emblem of the road to liberty. Bone may walk this road—she is, after all, too young to drive—but the roar of the cars reminds us of their pervasive role in determining her identity: as white trash, as victim, and, ultimately, as escapee and survivor.

The emblem of the North Star reconfirms the racializing of white trash identity that Allison plays with earlier in the novel. Several scholars have debated the extent to which "white trash" constitutes a racial category, and Allison's work has largely been read primarily through the lens of class and lesbianism.[1] Moira P. Baker notes, "In the novel Allison demonstrates how the construct 'white trash' creates not a racial but a class-defined Other" (120). Kelly L. Thomas also focuses on class issues, arguing, "Allison's poor whites don't subscribe to middle-class notions of progress, and hence, do not conform to societal norms regarding sexuality and citizenship as the epithet 'bastard' indicates" (168). Allison's characters do not conform to social norms of citizenship; rather, they reconfigure it as mobile. But part of what separates them from middle-class ideology is race, a concept to which Bone seems particularly attuned. When her aunt Alma temporarily leaves her husband and moves with her children into an apartment building also populated by an African-American family, Bone finds herself attracted to the young black girl but can't bring herself to express this feeling aloud. Later she accompanies her friend, Shannon Pearl, on a trip so that Shannon's father can discover new gospel talent and hears what she realizes is "real gospel" of "gut-shaking, deep-bellied, powerful voices" coming from an African-American church (169).

But Shannon's contemptuous claim that her daddy "don't handle niggers" causes Bone to realize that "the way Shannon said 'nigger' tore at me, the tone pitched exactly like the echoing sounds of Aunt Madeline [Daddy Glen's middle-class sister-in-law] sneering 'trash' when she thought I wasn't close enough to hear" (170). Bone astutely realizes the connection between *nigger* and *trash*, a connection made real by the car they drive. While visiting Daddy Glen's brothers, one a dentist and one a lawyer, she hears them pass judgment on the family vehicle: "Look at that car. Just like any nigger trash, getting something like that. . . . Her [Anney] and her kids sure go with that car" (102). It makes sense that the machine with the power to reconfigure citizenship also has the ability to re-race the white trash subject as black. Given this subtle identification with blackness, it makes sense that Bone imagines freedom from abuse as following the road northward, guided by the North Star. And given the role of the automobile in constructing both race and class, it makes sense that Bone imagines cars racing up that northern highway.

Allison's novel thus offers a fascinating picture of automotive citizenship—an unstable site that shapes an uneasy and contested sense of belonging. More fragile and less private than the home, the car reflects the instability of the relation between the individual and the state, particularly for those outside of the mainstream. What happens within the home generally stays private, although, as Berlant argues, by the 1980s those private values had become the stuff of public discourse and civic responsibility. But what happens within the car may classify you within the culture at large. The historian David Lewis points out that the city of Chicago, "by legislating against any 'indecent act in public' and defining sexual intercourse as 'indecent' and a car as a 'public place,' forbids lovemaking in cars" (124). Automotive space is public; the car, after all, wrecks easily, spilling one into the public sphere. Such wrecks reveal illegitimacy, class status, and sexual assault; in other words, car episodes have legal repercussions. And those repercussions remind Bone of her marginal status within the state. Even the emergency room doctor who urges her to speak up while treating earlier injuries fails to gain her trust: "He didn't know us, didn't know my mama or me" (114).

The sheriff who tries to investigate the rape finds her equally uncooperative: "Son of a bitch in his smug uniform could talk like Santa Claus, promise anything, but I was alone" (297).

By reminding Bone of her non-status in the eyes of the state—she is unknown and alone—these events challenge any notion of civic responsibility. If even an abused child cannot feel protection or support from the state, then any conventional notion of citizenship is a sham. Automotive citizenship may not protect Bone from abuse, but neither does it support her abusers as pillars of the community, the way Larry Cook is perceived in *A Thousand Acres;* legitimacy and legality are not necessarily aligned in this text. In fact, by making clear the extent to which conventional citizenship is privileged—limited to the white middle-class such as Daddy Glen's family and the manager of the Woolworth's who bans Bone from the store—automotive citizenship demands a reconceptualization of citizenship as a force. It opens a space between the public and private, one that reveals family secrets without either pathologizing or idealizing the family. This counters the move to ensconce conservative "family values" as the defining core of U.S. citizenship and leaves open the possibility for something more flexible, a concept of citizenship that recognizes more than national origin or class status, a concept of citizenship shaped by the automobile.

Ethnic Automotive Citizenship

Citizenship, of course, is a particularly contested domain for nonwhites. And as Peter Marsh and Peter Collett argue, those who have been devalued by mainstream culture often use cars as symbols of self-worth (107). But cars can also function as symbols of access to American identity and as bridges among dominant and nondominant cultures. Race, in particular, may be shaped or even constructed by the car, as demonstrated early in the century by Faulkner, Boyle, and Petry and, more recently, in Senna and Allison. Living as a nonwhite in America renders one particularly dependent on automobility. From Native American texts struggling against the power of white culture—often emblemized by cars—to immigrant novels in which characters seek to belong, the automobile shapes the definition of U.S. citizenship. The writers Leslie Marmon Silko, Cristina Garcia, Julia Alvarez, and Cynthia Kadohata all explore how nonwhites construct or resist automotive citizenship. The car allows one to negotiate between different cultural and ethnic spaces, to recycle the ultimate American icon, the automobile, into a vehicle that reflects diversity rather than dominance.

For Silko in *Ceremony* (1977) the car is a machine to be adapted to the Native American way of life, a vehicle that may emblemize white culture but is capable of assisting in a kind of hybrid identity. Old Betonie, the medicine man, lives in the hills over

Gallup, surveying the town dump. When questioned about why he remains there, he replies: "But see, this hogan was here first. Built long before the white people ever came. It is that town down there which is out of place. Not this old medicine man" (118). Betonie articulates a sense of location that both incorporates nationality and transcends it. "The 'locality' of national culture," notes Homi Bhabha, "is neither unified nor unitary in relation to itself, nor must it be seen simply as 'other' in relation to what is outside or beyond it. The boundary is Janus-faced and the problem of outside/inside must always itself be a process of hybridity" (4). By refusing to locate himself as "other," Betonie accepts his place as both inside and outside white culture. He says he is "comfortable" there, a concept the protagonist, Tayo, has difficulty accepting: "But the special meaning the old man had given to the English word [*comfortable*] was burned away by the glare of the sun on tin cans and broken glass, blinding reflections off the mirrors and chrome of the wrecked cars in the dump below" (117). For Tayo the "blinding reflections" off wrecked cars desecrates the landscape, reminding him once again of white theft of Native American lands. Thus, the car seems to represent a kind of citizenship he despises, that of white Americans who exploit the land. Unlike Betonie, who privileges hybridity, Tayo, also a half-breed, resists what white culture represents: garbage dumps and wrecked cars.

Later Tayo's friend Harley brags about the truck he has just acquired: "No money down! Pay the first of the month!" The catch being that they need to find him to obtain payment, and Harley has every intention of not being around when the bill collectors come calling. For him this maneuver helps to even up the balance sheet: "They owed it to us—we traded for some of the land they stole from us!" But the broken-down truck hardly compensates, as Helen Jean, the only woman present in the scene, quickly points out. "Gypped you again!" she says. "This thing isn't even worth a half acre!" (157). The attempt to exploit automobility as a means of asserting power falls pathetically short, partly because the form of resistance hinges upon exploiting white values: the monetary worth of a truck. Yet Silko complicates the situation more fully by gendering it: through Helen Jean who sees the emptiness of the truck's promise and, even more tellingly, through Harley's name, which evokes the most macho of automotive symbols, the Harley-Davidson motorcycle. This move reverses the custom in the auto industry of using Native American names for cars: Cherokee, Pontiac, Dakota, and so on. By claiming the Harley as a Native American name, Silko subtly challenges both the symbolic appropriation of Indian culture by car manufacturers and the implied masculinity of such culture. Harley's name reminds us that macho masculinity and macho cars, aligned with white dominance, are equally out of place in this Native American community.

Significantly, white culture in this book often arrives by car. The army recruiter, who rather grudgingly allows Tayo and his cousin Rocky to sign up in the increasingly tense days immediately prior to World War II, arrives in a government car and condescendingly lectures the two young men, "Now I know you boys love America as much as we do, but this is your big chance to *show* it!" (64). Their status is clear: they are marginal Americans at best, offered the opportunity to prove themselves true citizens by joining the army, the site in which men—and citizens—are made. But for them citizenship goes only as far as the uniform. Tayo remembers a white woman blessing them from a car, "but it was the uniform, not them, she blessed" (41). The American identity owned by the recruiter and the white woman—both in cars—is merely loaned to Native American soldiers as long as the war lasts, and Harley's truck does not extend access to it. Indian soldiers may celebrate success with white women during the war, but even there they often masquerade as Italians, thereby undermining Indian masculinity; they can attract women only by posing as non-Indians. White masculinity does not extend to Native Americans, regardless of whether they carry the names of white male sexual symbols such as the Harley. Car culture appears to reinforce white dominance.

Yet when Tayo is perceived as a threat and hunted by government people in cars, white automotive culture bogs down on Indian land: "Their Government cars will get stuck in sand and muddy places" (233). Moreover, the automotive threat is not restricted to whites; Tayo is in far greater danger from Pinky and his cohorts, the Laguna witches who want to stop him from completing his healing ceremony. And they, too, come by car. As they smash a tire iron against a car's hood, Tayo realizes that it was "the sound of witchery" (250). An integral element of U.S. government and Native American witchery, the automobile forms an uneasy link between the two, highlighting the interconnectedness of white and Native American culture and reminding us that cars can represent the forces of danger and oppression. Philip Deloria notes, "Symbolic systems surrounding Indians (nature, violence, primitivism, authenticity, indigeneity) and automobiles (speed, technological advance, independence, identity, progress) continue to evoke powerful points of both intersection and divergence" (141–43). Here the intersection of violence and technology indicates the extent to which cars have become integrated into Native American culture.

That integration is not always presented as evil; Tayo dismisses old Betonie's comfort in his space too easily. Betonie, in effect, recycles the automobile, cooking on "a grill he had salvaged from the front end of a wrecked car" (127). By refusing to dismiss the car as an alien machine in the garden, he opens a space for negotiating between technology and nature, between white and Native American, highlighting the hybrid identity that comes off as the hope for the future in this text, from Tayo himself to the hybrid spotted cattle. To incorporate the car into Native American culture is to practice a form of automotive citizenship, a way of negotiating around boundaries and accepting a fluid sense of belonging. As Tayo finally realizes: "He was not crazy; he had never been crazy. He had only seen and heard the world as it always was: no boundaries, only transitions through all distances and time" (246). The car helps to reshape and destabilize rigid conventions of who belongs. Tayo is integrated into his community, reconnected with his Native American heritage but also, finally, accommodating the world of technology: "In the distance he could hear big diesel trucks rumbling down Highway 66 past Laguna. The leaves of the big cottonwood tree had turned pale yellow; the first sunlight caught the tips of the leaves at the top of the old tree and made them bright gold. They had always been loved. He thought of her [both Ts'eh, the spirit of the sacred mountain, and his own mother] then; she had always loved him, she had never left him; she had always been there" (255). Here the sound of the trucks, rather than signifying witchery or white government mismanagement, meshes with the spirituality; neither is out of place as the automotive noise brings femininity to Native American car culture, reinforcing the fluid gender boundaries.

Thus, Silko fashions a world in which cars can coexist with nature, creating a space that allows for hybridity and a way of belonging that integrates elements of both Native and white culture. Deloria argues, "When Native drivers took the wheel, the world of technology became a landscape of dangerous intersections, with the likelihood of spectacular crashes between the expectations of the white drivers found on Automotive Avenue and the Native drivers who turned in from Indian Street" (168). In other words, the presence of Indians in cars destabilizes automotive culture, thereby reframing an American identity so dependent on automobility. In *Ceremony* Silko formulates an automotive citizenship at the intersection of traditional Native American culture and the car, a hybrid identity that blurs the boundaries of nature and technology, and male and female, a citizenry of "feminized" men such as Tayo, unafraid of love and emotion. By challenging the masculinity of car culture, she also challenges its position as an emblem of white culture. Just as gender functions less oppositionally, so cars can bridge diverse cultures to offer a space of automotive citizenship that foregrounds the intersection rather than the divergence of white and Native American identity.

Writers detailing immigrant conditions, such as Cristina Garcia and Julia Alvarez, similarly represent the car as integrally related to American identity. When Lourdes, the anti-Castro capitalist who left Cuba in the wake of the revolution, returns for a visit in Garcia's 1992 novel *Dreaming in Cuban,* she cries out to Cuban drivers: "You could have Cadillacs with leather interiors! Air conditioning! Automatic windows!" (221). In looking to find reasons for privileging U.S. citizenship above Cuban, it is no accident that Lourdes seizes upon cars to make her point. The Cadillac—the capitalist's dream car—epitomizes American cultural identity. Those, like Lourdes, who want to be a part of U.S. consumer capitalism, drive luxury cars. Alvarez links the car even more strongly to U.S. identity. In *How the García Girls Lost Their Accents* (1992) the adolescent Carla, a recent immigrant, finds herself lost among Americans: "It was hard for Carla to tell with Americans how old they were. They were like cars to her, identifiable by the color of their clothes and a general age group" (155–56). One could hardly ask for a more pointed observation on American identity being linked to the automobile. Carla strives to be accepted (or, at least, left alone) by the boys who "talked excitedly about Fords and Falcons and Corvairs and Plymouth Valiants." She "sometimes imagined herself being driven to school in a flashy red car the boys would admire. Except there was no one to drive her. Her immigrant father with his thick moustache and accent and three-piece suit would only bring her more ridicule. Her mother did not yet know how to drive. Even though Carla could imagine owning a very expensive car, she could not imagine her parents as different from what they were" (155).

Like Garcia's Lourdes, Carla recognizes expensive cars as signifying acceptance within the U.S. community. Interestingly, however, she cannot conceive of her immigrant parents as able to participate in automotive culture. Cars grant access to American culture; immigrant families do not. In some ways the car isolates Carla from her family, forcing a break between her Dominican and U.S. identities. Just as Allison suggests that family values may be filtered through automobility, Alvarez notes the extent to which cars may trump family as a means of shaping identity. But while Carla hopes that a fancy car will erase the difference between her own dark skin and the skin of the white boys who torment her—will make her "American" and therefore safe—the car cannot erase ethnicity. Further, it serves as a reminder of female vulnerability, associated with the boys who taunt and threaten her and a man who exposes himself to her from a car. Although she perceives her father as an unacceptable driver, her mother does not drive at all. Thus, the car represents not just American culture but also patriarchy—in both the United States and the Dominican Republic. The one make of car that Carla can identify is a black Volkswagen, the car of the secret police back home. The sexual threat of America and the threat of political oppression in the Dominican Republic both hinge upon cars, which may explain her fantasy of a flashy red car; only another car can challenge these attempts to keep her in her place and grant her safe access from home to school, from private to public.

Focusing on the car—replacing the black VW of the Dominican Republic with the flashy red car identified with her American experience—seems a doable task, a way of conceptualizing American citizenship that is not beyond reach. By identifying national identity with specific cars, she reveals the extent to which cars shape belonging.

To be American rather than Dominican is to drive a red car—and to sacrifice her Dominican parents. Yet by focusing on the car, she forms a link to her Dominican past, also defined by cars. This realization allows Carla to begin to develop her auto-motive citizenship, an identity that incorporates her Dominican past with her American present. While struggling with class differences, language differences, and different social experiences after their move to America, the García girls can look to the car as a tangible means of negotiating a hybrid identity, of erasing boundaries and desta-bilizing legal citizenship.

Cynthia Kadohata, in *The Floating World* (1989), explores this immigrant space more fully through the Japanese-American Osaka family, who perform migrant la-bor, though they may spend several years in any one place. Olivia, the narrator, says, "My own earliest memories were of pictures from a car window—telephone wires illuminated by streetlamps, factories outlined against a still, sunless sky—pictures of one world fading as another took its place" (55). Her origin seems to be the auto-mobile, as the family travels the interstices of a world hostile to those of Japanese descent: America in the 1950s. The car replaces the larger community, as the family seeks out other Japanese families and shuns involvement with the white mainstream culture. But the culture of the car eerily intersects with Japanese culture: "We were traveling then in what she [the grandmother] called ukiyo, the floating world. The floating world was the gas station attendants, restaurants, and jobs we depended on, the motel towns floating in the middle of fields and mountains. In old Japan, ukiyo meant the districts full of brothels, teahouses, and public baths, but it also referred to change and the pleasures and loneliness change brings" (2–3). The automobile recon-figures the Japanese floating world in America, offering a location for a potential Japanese-American identity and automotive citizenship.

The American floating world, constructed by the automobile, is a world of mobil-ity. Gas stations, restaurants, and hotels replace the teahouses, brothels, and public baths of Japan. In other words, a world of ritual and ceremony gives way to migrancy, to change. And despite its grounding in the American landscape, this floating world also gives way to unreality. After observing an accident in which a woman may have been killed, Olivia muses: "To me the accident was already becoming something that had happened, *just* something that had happened, the way a fire we might pass on the road was just a fire we passed on the road, even though it was someone's house or farm burning. It was just part of Obâsan's floating world" (37). Later, as a young adult driving around servicing vending machines, she notes: "There were no lights along the highway and sometimes it was hard to tell we were in a real car, because the scenery on either side changed hardly at all. It was like one of those pretend cars in arcades, where you have a seat and a steering wheel" (154). The world of automobility becomes a floating world, a space between Japanese and American culture, between material reality and the imagination. Given that Olivia's earliest memories come from this world, her sense of citizenship seems to belong there rather than in any formulation of nationalism.

While Olivia associates this automotive citizenship with the imagination, issues of Asian-American citizenship have a complex legal history. Brook Thomas has explored the impact of the 1898 Supreme Court case, *United States v. Wong Kim Ark,* on citizenship because the case established that citizenship could not be determined by "racial descent" but was, rather, based on the nation of one's own birth (690–91). The case opened up considerable possibilities for Chinese Americans who had previously been denied American citizenship, regardless of having been born on American soil, and Thomas goes on to examine Maxine Hong Kingston's 1980 novel *China Men* in light of its repercussions. We can see some of the same issues at work in Kadohata's text. Racial descent may not determine citizenship, but for Japanese Americans in the 1950s it certainly worked as a form of exclusion from participation in the American public sphere. By creating an alternative form of automotive citizenship, Kadohata's characters construct a world that accommodates nonwhite ethnicity but is shaped by one of the most popular symbols of America: the car.

Thus, this floating automotive world, firmly grounded on American soil, allows those born in the United States of Asian parents to claim the citizenship promised by the Supreme Court—but certainly not readily available to them ten years earlier, during the infamous roundups of World War II. Indeed, that immediate history may help to illuminate further why this must be a floating world: people of Japanese descent cannot situate their American identity too firmly; the floating world, however, allows them to float—by car—amid the American landscape, claiming U.S. citizenship but also retaining their Japanese heritage. Published in 1989, by which time the success of the Japanese automobile industry had been well established, this novel reminds us that the car, in fact, serves as the perfect vehicle to bring together Japanese and American identity. The automotive floating world allows for global positioning within a local context, reshaping America's sense of belonging to the world at large and, at the same time, reemphasizing American identity. As Melani McAlister reminds us, "Highlighting alternative images of (transnational) community . . . does not in itself undo the nation; it simply reminds us again that the nation is a modern human artifact" (423). One of the significant defining artifacts of the United States as a nation is the car, which offers alternative possibilities for defining national identity.

Viramontes: Automotive Citizenship in the Borderlands

About a third of the way through Helena Viramontes's *Under the Feet of Jesus* (1995), a novel about migrant workers in California, Estrella, the young Chicana protagonist, stops to watch a baseball game on her way home from the fields. But this all-American scene is quickly disrupted when the stadium lights throw her into a panic: "She startled when the sheets of high-powered lights beamed on the playing field like headlights of cars, blinding her. . . . The border patrol, she thought, and she tried to remember which side she was on and which side of the wire mesh she was safe in. . . . Where was home?" (59–60). In many ways the scene encapsulates the aura of the border culture described in 1989 by Gloria Anzaldúa in *Borderlands / La Frontera*—and by many critics since then—with the image of the fence, the fear, and, most particularly, the confusion over which side constitutes home. Yet it is telling that the lights that inspire Estrella's fear and uncertainty are likened to car headlights. For in this border culture it is the car that determines one's origins. In a novel in which the characters—both U.S. citizens and legal residents with green cards—live with the fear and appalling conditions associated with undocumented workers, home loses its associations with national origin. Ultimately, it doesn't matter where people were born or that the children's birth certificates, documenting their U.S. citizenship, are hidden under the feet of the Jesus shrine that the mother sets up in every ramshackle hut they inhabit; their origin is neither Mexico nor the United States. It is the car.

In this novel Viramontes adapts the border culture of Anzaldúa but with a significant twist; the borderland becomes the land of automobility. Anzaldúa presents the borderland as "a thin edge of / barbwire," a "vague and undetermined place created by the emotional residue of an unnatural boundary" (3). She historicizes that place in considerable detail, linking it to the history of American imperialism, Anglo oppression, broken treaties, and current U.S. agribusiness practices and immigration policies. While Anzaldúa certainly does not offer the borderland as some sort of idyllic location, a place of uncontested liberation and flexible power, she does argue—by insisting on its Mexican origin—that it provides a grounding for Chicana identity, an identity constructed by something beyond place and nationality: "Deep in our hearts we believe that being Mexican has nothing to do with which country one lives in. Being Mexican is a state of soul—not one of mind, not one of citizenship. Neither eagle nor serpent, but both. And like the ocean, neither animal respects borders" (62). More recent scholarship has built upon Anzaldúa's work, with critics such as Debra Castillo and María Socorro Tabuenca Córdoba calling attention to the need to look more closely at the specifics of the border, particularly the Mexican side. Yet Viramontes's novel echoes Anzaldúa's more metaphoric—and U.S.-focused—construction of border culture. It turns out to be remarkably difficult to date and place the novel, both intensifying the sense of this book taking place in a world between Mexico and the United States but also reminding us of its lack of grounding in place and nationality.

Automotive identity bears a marked similarity to the borderland culture that Anzaldúa describes. It unfixes place, erases boundaries, and moves between multiple locations. While obviously it is not exclusive to the United States, it holds particular power within U.S. culture. Cecelia Tichi refers the American tradition "exalting the kinetic, the action-oriented. The *real* Americans, this tradition says, are mobile, migratory, questing, surging, driving, continually on the move" (*Country Music* 72). Certainly, the migrant workers of this novel thus qualify as real Americans, though in a slightly different spirit. Automobility in the borderland may be a state of soul rather than mind or citizenship. But it's also very much a place of the body. In her attention to the material conditions of the Chicana body, Viramontes ensures that the borderland space does not become merely symbolic. Manuel Luis Martinez challenges the valorization of a symbolic borderland, arguing that it dismisses "the importance of 'place' and 'citizenship' in a nationally defined entity that permanently denies especially the (undocumented) migrant arrival" (54). By insisting on a kind of mobile identity, Martinez claims, we keep the migrants moving, often ignoring the material conditions of migrants' lives, many of whom seek inclusion as U.S. citizens: "In rightfully challenging non-democratic, non-participatory versions of the nation-state, we have endorsed the creation of an immaterial, so-called 'hybrid' culture rather than an inclusive national culture that calls for social justice" (57).

But Viramontes, in constructing automotive citizenship, acknowledges both the desire to belong and the flexibility of borderland identity, both the material reality of the automobile and the mobility—literal and theoretical—it represents. The border, we are reminded by Sandra Cisneros in her 2002 novel *Caramelo,* is infested by cars, a veritable traffic jam: "Not like on the Triple A atlas from orange to pink, but at a stoplight in a rippled heat and a dizzy gasoline stink, the United States ends all at once, a tangled shove of red lights from cars and trucks waiting their turn to get past the bridge. Miles and miles" (16). Chicana writers are well aware of the role cars play in establishing the borderland, cars that are driven by human bodies. Cars serve as ethnic markers, particularly in the Chicano community, where the low-rider has become a work of art.[2] Rather than erasing the border, cars proclaim its presence—along with the presence of Chicano workers and drivers.

In her careful attention to the characters' bodies Viramontes provides a grounding that is lacking in the uncertain time and place. And the focus on the body links identity to the car in this text. Automotive citizenship means that one comes from a place of constant movement, one's body always dependent on automotive technology. *Under the Feet of Jesus* opens with the family's arrival at another labor site in their dilapidated station wagon: "It was always a question of work, and work depended on the harvest, the car running, their health, the conditions of the road, how long the money held out, and the weather, which meant they could depend on nothing" (4). Bodily health and the health of the car are equated, and the car comes first, the state of their bodies reflected in the knocking skipping engine, the aging car battery. Similar concerns are expressed in an earlier text of migrancy, John Steinbeck's *Grapes of Wrath,* in which the attention that used to be paid to farming is now devoted to the car that will make or break the family's existence: "Eyes watched the tires, ears listened to the clattering motors, and minds struggled with oil, with gasoline, with the thinning rubber between air and road. Then a broken gear was tragedy" (268). In both instances the importance of the car determines the family's place: they are migrants, and not much seems to have changed in the roughly fifty years between the two novels. Steinbeck's Okies, however, insist upon their American identity throughout their ordeals: "We ain't foreign. Seven generations back Americans" (317). By the latter part of the twentieth century this aggressively asserted American identity proves more elusive for the Chicano migrant workers of Viramontes's text. And when people are denied a place in American culture, then car culture is all that is left to determine identity and even origin.

Despite the largely Chicano focus of the novel, Anzaldúa's insistence on Mexican identity is largely absent, as are specific references to Mexico as place. Even the baseball scene cited earlier does not specify which side of the border constitutes safety. Indeed, this novel resists looking to Mexico as the lost home of the largely Chicano workers; only Perfecto, the elderly stepfather, longs to return "before home became so distant, he wouldn't be able to remember his way back" (83). While there are questions about which side of the border one is on, Mexico exists primarily in Perfecto's imagination. Estrella, born in the United States, appears to lack any sense of Mexican identity, and her friend Alejo identifies himself as a fifth-generation Texan, the spelling bee champ of Hidalgo County. Furthermore, although the main characters are Chicano, Viramontes also presents white Anglo and Japanese migrant workers. The inhabitants of this borderland, then, do not necessarily claim a common Chicano/Mexican origin; what links them is their migrancy, their dependence on automotive culture, from the rusty cars they drive following the harvests to the trucks that transport them from camps to fields. Thus, the automobile—rather than nationality or even ethnicity—becomes the constant factor in determining identity.

This text, with its poetic prose and dislocated narrative form, traces a few weeks, with some flashbacks, in the lives of a family of Chicano workers and a young man,

Alejo, who is attracted to Estrella, the eldest daughter. But despite the claims on the back cover ("Within Estrella, seeds of growth and change are stirring. And in the arms of Alejo, they burst into a full, fierce flower as she tastes the joy and pain of first love"), this is not a teenage love story. Just as Dorothy Allison resists portraying the self-sacrificing mother, Viramontes understands that teenage girls may have more pressing concerns than puppy love; while Estrella may enjoy Alejo's company, her greater concern is the survival of her family. Rather, this novel emphasizes the abhorrent conditions of migrant laborers and the difficulties of familial and personal survival in such a life. What's striking about this family group is their rootlessness; they are never referred to by family name and never—except for the elderly Perfecto—identify any place as home. Given how often their place of residence changes, it becomes clear that the car is the only constant. Although they rarely literally live in the car, it constitutes the only home they keep coming back to. As Steinbeck puts it, the car "was the new hearth, the living center of the family" (*Grapes* 136).

In fact, Petra, the mother, looks longingly at a man driving a fancy car, imagining that his car also serves as home, albeit a much more luxurious one: "The white plush carpeting was so white, it was obvious no one ate in the car. She envied the car, then envied the landlord of the car who could travel from one splat dot [on the map] to another. She thought him a man who knew his neighbors well, who returned to the same bed, who could tell where the schools and where the stores were, and where the Nescafé coffee jars in the stores were located, and payday always came at the end of the week" (105). Petra does not fantasize about where the man may live; rather, she sees him as the "landlord" of the car, privileging car over house. It is the car that gives him a sense of belonging, that constructs who he is: a man familiar with his surroundings, a man who belongs. As Paula M. L. Moya points out, "The Bermuda man functions in the novel as a stand-in for the middle-class American whose social location inhibits his apprehension of the real economic relations in which he exists" (194). His economic status is defined, in this passage, by his car. While one assumes the man is Anglo, the only specific information we receive about him is that he drives a nice car and thus is clearly a U.S. citizen, in a cultural if not legal sense. Petra and her family, legal citizens, lack such cultural recognition.

This automotive exclusion is not limited to nonwhites. Bob DuBois, of Russell Banks's *Continental Drift*, experiences a similar realization of his place within the dominant affluent culture. Seeing a white convertible, he "gets out of his car and strolls around it and for a few seconds admires the Chrysler, standing next to it while he finishes off the can of Schlitz, rubbing his tee-shirted belly and examining the rolled and pleated red leather upholstery, which smells like nothing but itself and reminds Bob of polished wood, Irish tweed, gleaming brass. Glancing at his own face in the tear-shaped outside mirror, Bob suddenly sees himself as he must look from inside the store, a man in work clothes guzzling beer and drooling over someone else's luck" (238–39). Fancy cars seem to convey a particular sense of belonging and acceptance, one largely denied to nonwhites and working-class white men. While both Petra and Bob depend upon functional cars for their livelihood, neither participates in the luxuries of automotive culture. Unlike expensive cars, however, automotive citizenship is accessible to the nonelite.

These passages remind us that cars do not simply signify mobility; rather, as in the work of Alvarez and Allison, they serve as markers of one's place within U.S. culture. The space they offer these migrant workers marks not just their economic status but their very identities. "Chicanas," says Mary Pat Brady, "write with a sense of urgency about the power of space, about its (in)clement capacity to direct and contort opportunities, hopes, lives. They write also with a sense of urgency about the need to contest such power, to counter it with alternative spatial configurations, ontologies, and geneologies" (9). The car is one such alternative, not because it offers any kind of escape from poverty, discrimination, and migrancy but because it suggests a way of conceptualizing identity that breaks from the U.S.-Mexican dichotomy. It provides a different genealogy. It's not that automobility constitutes a "better" alternative but that it intertwines with ethnicity to reflect a culture in which people are becoming increasingly *dis*placed. In the novel only the elderly have any sense of home or place; the young, to all intents and purposes, come from the car. Sonia Saldívar-Hull notes of Viramontes's stories that they "are not a quest for origins; what the *historias* offer are alternatives" (131). In *Under the Feet of Jesus* that exploration of alternatives hinges upon the car as origin.

The car is crucial not only because it functions as a kind of national origin but also because it operates through the very bodies of the migrants themselves. Alejo, who wants to be a geologist, describes for Estrella where oil comes from: "The bones lay in the seabed for millions of years. That's how it was. Makes sense, don't it, bones becoming tar oil?" (87). He goes on to relate the finding of the bones of a young girl in the La Brea Tar Pits in Los Angeles. Estrella astutely puts this information together later in the novel when a nurse at a clinic takes all their money to look at Alejo, who has been poisoned by pesticides, and simply tells them to take him to a hospital, an impossible task because they now lack money for gas: "She remembered the tar pits. Energy money, the fossilized bones of energy matter. How bones made oil and oil made gasoline. The oil was made from their bones, and it was their bones that kept the nurse's car from not halting on some highway. . . . Their bones. Why couldn't the nurse see that? Estrella had figured it out: the nurse owed *them* as much as they owed her" (148). Moya notes that "the trope of fuel" functions as "a metaphor for the situation of migrant farmworkers," whose labor in picking produce "literally provide[s] the 'fuel' that keeps our bodies going" (195). But *fuel* encompasses gas as well as food, as Estrella's realization of the power of her own body grants her automotive power and assigns her a place in the world.[3]

Their bones fuel the car, putting them in control of automobile culture. This is a marked shift from *Grapes of Wrath*. While the car is equally important to the Joad family in that novel, they perceive gas as money and strain to be able to keep the car running by sacrificing even food. But they do not go so far as to identify themselves as fuel. Joyce Carol Oates, in *them*, anticipates the perception of the human body as fuel when Jules worries "that he was like the gas in the car's tank. He, himself, was the gas. He had to keep them going; he was running down; he had to be juiced up" (285). Yet for Oates, Jules's initial identification with gas is figurative; she carefully phrases it as a simile. Jules is *like* gas. Estrella, however, literally sees her body as oil, the stuff of which gas is made. Viramontes, more than Oates, grounds automotive energy in the human body. And it is this realization that inspires Estrella to demand their money back and, when the nurse resists, to begin smashing items on the nurse's desk with a crowbar. It is this realization of her automotive citizenship that moves her to an activist consciousness: "You talk and talk and talk to them and they ignore you. But you pick up a crowbar and break the pictures of their children, and all of a sudden they listen real fast" (151). Petra has often tried to instill in Estrella a sense of pride—"Don't run scared. You stay there and look them in the eye. Don't let them make you feel you did a crime for picking the vegetables they'll be eating for dinner" (63)—but picking vegetables does not touch her in the way her realization of automotive power does. The close of the novel finds Estrella, presumably budding into an activist, symbolically replacing the now-broken statue of Jesus as she stands atop a barn, the shakes crunching "beneath her bare feet like the serpent under the feet of Jesus" (175).

Even the imagery links Estrella, Jesus, and the car. The doily that sits under the Jesus shrine, covering the mass of documents—birth and marriage certificates, identity cards, certificates of baptism—is crocheted in "perfect little diamonds" by Petra's grandmother (165). When Estrella stands on the barn at the end of the novel, looking up, she is "stunned by the diamonds" (175). Seeing the stars as diamonds connects Estrella—Star—with diamonds, and thus with Jesus, and also with the creativity of her great-grandmother, who made the diamond doily. But the first reference to diamonds in the book occurs when Estrella lies underneath a truck for shade and notices "the way the tire tread actually had a lacy diamond pattern like the scarf doily the mother spread under Jesucristo" (87). In bringing together the automobile, Jesus, and Estrella through her diamond imagery, Viramontes intertwines one of the most prominent aspects of Chicana identity, Jesus, with an equally powerful symbol, the car. Thus, she constructs a protagonist whose ethnicity and origin intersect with automotive culture. Rather than erasing Chicana identity, she illustrates that ethnicity is also shaped by the car, particularly in a text in which Mexico plays a more metaphoric than actual role.

Indeed, the place itself is very difficult to locate, though it appears to be southern California. The year is equally difficult to pin down. The main clues come through Perfecto, who says he's seventy-three but is not sure of the year of his birth, except for the year 1917, which comes to him in a dream. This, if accurate, places the text roughly in 1990. But there are no references to politics, to labor organizing, nothing that would ground the novel in a specific time and place. Estrella at one point is given a leaflet with an eagle on it, probably the firebird emblem of the United Farm Workers union, but she never appears to have any contact with the group.[4] It is, to echo Anzaldúa, "a vague and undetermined place." The brand names all refer to goods that have been around for years: Spam, Nescafé coffee, and Quaker Oats. This lack of specificity ungrounds the novel, making the car an even more appropriate emblem of identity. And the car contributes to the unfixed nature of the text. Viramontes plays with car names to both ground and unground automobile culture as a defining characteristic. She identifies the luxury car as a "Bermuda," a model that has never existed. The family drives a "Chevy Capri," another anomaly; there has been a Chevy Caprice and a Lincoln Capri but no Chevy Capri. While it's possible she got mixed up amid the multitudinous car makes and models, the fact that both named vehicles are slightly off suggests a deliberate purpose. It suggests a borderland car culture in which established norms and assumptions don't quite apply, in which things—and people—can be reclassified at will, in which definitions may be arbitrary.

The border, says Brady, "works in part as an abjection machine—transforming people into 'aliens,' 'illegals,' 'wetbacks,' or 'undocumented,' and thereby rendering them unintelligible (and unintelligent), ontologically impossible, outside the real and the human" (50). Viramontes plays on this trope, first by appearing to accede to it. Her characters assume an illegal status, apparently unaware of what services they may be entitled to as U.S. citizens. In so doing, she illustrates the power of the abjection machine, as the border can turn all who reside near it into aliens. She further problematizes this alien status by aligning them with the nonhuman: oil. If they are the stuff of which gas is made, their human identity becomes all the more questionable. In fact, this association plays into discourses that reduce workers' bodies to tools. Anne Shea, examining the construction of narratives around farmworkers' bodies, cites a farmworker's claim, "We need to understand that migrants are not machines" (qtd. in Shea 135). Yet Viramontes identifies them not with machines but with what makes the machines run. It is precisely in this association between the body and oil that Estrella discovers her human potential. She is reborn, both inside and outside "the real and the human," in the borderland between the human and the automobile. In this way she erases her own abjection, for the abject terrifies us by reminding us of the fragile borders of the human body, but Estrella embraces that non-boundary, seeing it as her strength and entitlement.

None of this, of course, has much "real" impact in the book. Estrella does retrieve the money from the nurse, carefully taking only the exact amount they had paid. But she must dump Alejo at the hospital, unable to stay and comfort him because "the car ran outside, the white fumes rising from the exhaust pipe and the precious gasoline burned" (169). There is not enough gas, not enough of Estrella, to go around. In highlighting the precious gas, Viramontes also acknowledges Estrella's value: higher than diamonds, higher than oil. It may not be enough—at least not yet—but we do see a new direction opening up at the end of the book in which Estrella will go on to fuel a move toward agency and awareness.

The border culture Viramontes presents in this novel is subtly imbricated with automobility. By weaving the car so intricately into her text, she recasts origin and ethnicity. Certainly, the book, with its combination of Spanish and English language and its references to physical appearances commonly associated with Chicanos—to the tortillas they eat, to the labor they perform, to various cultural practices—functions as an overtly Chicana text. But that construction is also shaped by car culture. Identifying these characters as automotive citizens both places them within and excludes them from mainstream U.S. culture, but it also acknowledges their necessity within a nation dependent on automobiles. Gas is made of their bones. Thomas says of Maxine Hong Kinsgton's novel *China Men*, it "invites a rethinking of American identity by responding to the standard question—What makes an American—with the question—who made America? To answer that question is to alter the myth of the country's founding fathers by reminding us of the material, as well as the political, making of the country" (709). *Under the Feet of Jesus* might ask, Who feeds and fuels America?—reminding us of what it takes to sustain a nation. As necessary to the United States as the food migrant workers harvest are the cars they fuel. The car serves as both the origin and the product of migrants' lives.

In supplementing national identity with automotive identity, Viramontes opens another road from the borderland into the mainstream and so begins to crack open what Anzaldúa terms "a closed country" (11). After all, nothing is as ubiquitous as the car, both nationally and globally. And yet by claiming automobility as integral to the borderland, she both denies the demarcation of the borderland and also asserts its power in fueling Anglo society. Just as significantly, she highlights the power of the bodies of those who inhabit the borderland. Shea notes: "Through explicit and implicit juxtaposition against the citizen/owner, the laborer comes to be conceptualized and categorized as a non-citizen. To oppose citizenship and 'non-citizen labor' discursively and ideologically defines laborers as outside of the rights of citizenship, as outside of national definition" (130). But by aligning her characters with automotive citizenship, Viramontes works to erase that exclusion and drives the Chicana body into U.S. consciousness and cultural identity.

Automotive citizenship operates in borderlands of class, race, and geography, eliding the divisions between public and private, between reality and imagination. As such, it helps to break down the hierarchies that so often define women as lesser. It may not "liberate" individual women; Bone, after all, suffers multiple tragedies in cars, and Estrella's bones are metaphorically fed into gas tanks. Carla is terrorized by boys who talk about cars and by a man who exposes himself to her in a car, and cars pollute the land created by Thought Woman in *Ceremony*. But by challenging the ways traditional citizenship is constructed, automotive citizenship suggests a reformulation of American identity that is more flexible and more mobile, an unstable site that accommodates greater diversity and change. Henry Ford perceived the automobile as a great leveler, erasing social distinctions and bringing the rural population in touch with the urban centers. While his optimism may not have been borne out, American women writers have forced automobile culture to acknowledge women's presence, a vital step in achieving full civic recognition. Given America's dependence on the car, women's exercise of automotive citizenship has ensured that they will retain a place in the culture of the nation.

The early promise of the automobile, that it would reshape American society, has been more powerfully fulfilled than could have been anticipated. The day is not far off when a driver's license may be a prerequisite to voting, a twenty-first-century revision of the poll tax. While this is clearly a troubling political move, it does reflect the very American assumption that anyone who matters will have to be a part of the culture of automobility. We can now see that women such as Alice Ramsey, Edith Wharton, and Gertrude Stein were right: women did need to assert their rights to the motor car as a means of claiming their American identity. The early-twentieth-century suffragists who employed the car as an integral element of their campaign helped to set the standard for American citizenship: it belongs to those on wheels. Women writers, with their encyclopedic exploration of the automobile, have played their part in forming a society so focused on cars—a society, they were well aware, that could accommodate women and women's concerns.

From establishing women's driving abilities to facilitating women's relationship to automotive technology to refiguring maternity to reshaping female agency to dislocating domesticity to reformulating American citizenship, American women writers have illustrated the extent to which cars enhance women's participation in the nation. While cars, driving conditions, and women's concerns certainly changed over the course of the century, what has remained constant is that driving women have written the twentieth-century American literary tradition.

Writing behind the Wheel

There have been a lot of automotive changes over the past one hundred years; the "merry Oldsmobile" of the first decade of the twentieth century evolved into a negative—"not your father's Oldsmobile"—and then into extinction, along with many once-familiar names and brands. Time has diminished the dizzying excitement inherent in early-twentieth-century automotive culture as more and more people take their cars for granted and are less inclined to view them in Henry Ford's messianic light. But even amid the explosion of technological devices in the late twentieth and early twenty-first centuries, from computers to cell phones to iPods, the automobile continues to function as an integral part of American culture and American identity. Indeed, the automobile's power to convey life or death was literally driven home by Hurricane Katrina in the fall of 2005: those with cars were largely able to flee the area (albeit not easily, given traffic snarls and empty fuel tanks), while those without cars were most likely to be left behind either to die or to endure days of agonizing hardship waiting for help to arrive. Cars determine our very existence in American society.

Clearly, cars still rule. The auto industry itself has long been the mainstay of the American economy. Even as its power wanes—as globalization and a turbulent oil market take their toll—it remains a major benchmark of the American business landscape. As the twenty-first century gets under way, the difficulties facing the industry reflect the concerns confronting the entire business community. Long known for their generous benefits packages, the auto companies are now staggering under the load of rising health care costs. Toyota's 2005 decision to locate a new plant in Ontario rather than the United States was partly due to the fact that Canada has a national health care system, which alleviates the company's burden in paying for the coverage (see Krugman). Certainly, the auto industry is not alone in being caught by rising costs, but its visibility and dominance command attention when confronting changing business conditions. As the auto industry goes, so goes the nation. Thus, when the credit rating firm Standard & Poor's announced a downgrade of Ford and General Motors debt ratings to junk bond status in May 2005, the impact on the market was termed "a Richter-scale event" by Edward B. Marrinan, top investment-grade strategist at J. P. Morgan (qtd. in Fuerbringer). But the market and the public do not always see eye to eye. What does this "Richter-scale event" mean to the driving public? The answer, in terms of being able to purchase cars, is "not much." In these days of globalization the American automobile industry may be struggling to compete against foreign automakers, but people continue to buy—and depend upon—cars. There is disagreement about whether America's love affair with the car is over, but more than one hundred years after its birth the automobile has become ever more entrenched in American culture. W.E.B. DuBois perceptively anticipated that the problem of the twentieth century would be the problem of the color line; the technology of the twentieth century clearly was the automobile. As such, it is inextricably intertwined with the color line and, even more solidly, with the gender line.

That automotive gender line, however, is less stable than it used to be, as the popular image of a man working on his car has become largely obsolete. The increased computerization of the machinery means that to maintain a car or make repairs one needs sophisticated technology that is unavailable even in basic gas stations, let alone home garages. Calls to the radio show *Car Talk* generate very few questions about *how* to fix something; rather, most callers simply want to get an idea of what their problem is before approaching a mechanic so that they feel less likely to be taken for the proverbial ride. The disengagement from the actual workings of the vehicle leads, in a strange twist of events, to a somewhat more level playing field between men and women and their cars. Whereas women have always been assumed to be ignorant in such matters, men now find themselves similarly positioned on the margins. Evan Watkins, exploring the rhetoric in automobile maintenance manuals, notes that the content in car magazines has shifted from technical instructions to advice on detailing (155), which he aligns with "'traditionally' sanctioned women's work" (158). Even car repair, long the province of men, has become feminized, leading, not surprisingly, to what Watkins identifies as "a thematic of loss no less pervasive than in any nostalgia production. All the wealth of detailed information made available can hardly disguise a sense in which with newer cars there is, again, relatively very little that actually can be done at home" (156). Intimate knowledge of a car's inner working is largely a thing of the past. Few people, men or women, delve under their hoods to understand what makes their cars run—or fail to do so.

Increased automotive gender equality, then, encompasses a detachment from the car itself. This detachment, however, is primarily mechanical; most of us may no longer be able to fix our cars, but we remain largely dependent upon them and often emotionally attached to them. Over the years, when I describe my interest in the subject of women and cars, my listener's eyes light up, and I am greeted with stories about people's relationships with cars, especially from women. We may not fully understand how an automobile works, but we certainly understand what it means. For American women the car has meant autonomy, agency, power, and danger. It has both enhanced and undermined any notion of female space, providing escape from the home and expanding the boundaries of the home so that it often seems as if there is no escape; cars may script women's lives as surely as houses have done.

Few objects have done more to transform women's place in American culture than the automobile. And women, for their part, are leaving their mark on automobile culture and the automobile industry. In their insistence on taking, and keeping, the wheel, they have inspired a range of automotive developments, from enclosed cars and electric starters to power brakes, automatic transmissions, and safety features. Most interesting, though there are certainly features that women want in cars, what they apparently don't want is a car designed for them alone. Most such experiments have met with marked failure, the most famous being the 1955 Dodge La Femme, which came with matching accessories, including a rain cape, shoulder bag, compact, and lipstick. Throughout the 1950s one could buy clothes to match the upholstery of the Ford Victoria. As Ford's advertising copy announced, "It's a man's world, but woman runs it— and drives about it in this chic new coat—one of a collection called Motor-Mates" (Kudner Agency). The "man's world," apparently, would be off-limits to women not dressed in Motor-Mate clothes. The mind-set still prevails; in 2005 Ford began marketing a line of nail polish to match the paint colors of the Mustang (Schneider). While the development of the minivan seems to have met with greater success in terms of appealing to women, the manager of Chrysler's minivan operations, Ralph A. Sarotte, complains: "Someone put the 'soccer mom' title on it. We could sell twice as many minivans in this market if it were not for the image thing" (qtd. in Bradsher 109). The message is clear: not only do women, according to the industry, require special accommodation—generally linked to appearance—but once a car becomes a "woman's car," it loses much of its appeal.

If automakers try so hard to market to women, why does the taint of femininity poison a car's image? One answer may lie in men's apparent need to insist upon a manly car, to resist the implied emasculation of driving an effeminate car. Once minivans became mom cars, the industry produced SUVs. Then, after women began buying SUVs in record numbers, GM's Hummers hit the streets. Tough cars, one of the last bastions of male privilege, are no longer able to keep women at bay. Having gained the driver's seat, women are refusing to respect automotive gender boundaries. Ford may have hoped that the car would make a woman more the woman, but many of today's women deny that their vehicles reflect their gender. Nevertheless, some differences remain. Women, it might be said, refuse to treat the car with the deference that many men seem to feel it deserves. A *New York Times* article in 2002 claims that women executives, unlike men, rarely buy fancy cars to celebrate their success, being more inclined to indulge in expensive jewelry or spa treatments. "Driving a car is something you do every day," says Avis Yates, president of an information technology management company and a former Exxon executive. "A spa is something you do once a year" (qtd. in Stanley D1). Her remark reinforces women's attitude toward the automobile for the past century; for women a car tends to be less an extension of the self and more a tool, something to enhance power and agency rather than to serve primarily as a visible emblem of identity. And yet, while women may resist bowing down to the altar of automobility, they certainly have learned to take advantage of what it has to offer— in life and in literature.

American fiction reflects and shapes the dynamics between women and cars. The intersection of the two offers a position from which to explore the gradual shifts in how gender, technology, agency, autonomy, and domesticity have been articulated throughout the past hundred years. One sees not so much a gap between women's and men's presentation of the car as a twist, a claiming of automobility that resonates particularly strongly in women's novels. Just as "real" women do not generally perceive themselves as inhabiting a separate sphere within car culture, so women writers present their characters as very much a part of automobility. What we gain by considering a tradition of women's automotive fiction is precisely this realization of inclusion, a recognition of women's participation in automobile culture. For women the car does indeed help to level the playing field, to provide the power and mobility once largely restricted to men. Janet Guthrie, the first woman to race at the Indianapolis 500, responded to the question of whether a woman can compete with a man by saying: "I *drive* the car. I don't carry it" (qtd. in Dauphinais & Gareffa 138). This is certainly not to say that driving does not require skill, but, as Alice Huyler Ramsey declared in the first decade of the last century, such skill is accessible to women, a claim Danica Patrick clearly validated with her impressive 2005 rookie season on the Indy Car circuit.

By taking the wheel, women declared themselves full citizens of twentieth-century American culture. We cannot hope to grasp the significance of the car's influence without exploring its influence on half of the American population. Women writers, in documenting women's automobility, preserve and shape that automotive access. One of the most telling discoveries of this project has been the sheer number of writers employing cars in such a wide assortment of situations and with such a diverse range of meaning. Reading for the intersection of women and cars reshapes our understanding of twentieth-century women's fiction as fully cognizant of automobile culture and of its import to American culture. When we look at how women novelists have incorporated the car into their fiction, we can more easily see that women's fiction recognizes and produces new forms of women's mobility, agency, and identity. There is no single tradition or even map of twentieth-century American fiction, yet amid the many intersecting and diverging threads, the automobile emerges as a very important site of material and symbolic meaning, reminding us of the recasting of female power in the automotive age. Women's novels capture the reformulation of home, domesticity, place, and belonging, at least partially initiated by the development of the automobile.

Elaine Showalter has noted, however, that American women's writing has become so widely disseminated that it has ceased to be "a literature of our own" (21). The automobile's literary influence is certainly not restricted to female writers. Just as the automobile industry has become global, so has American fiction ranged in many diverse directions. Marc Chénetier has noted that the "mainstream" in recent American fiction "seems to be perpetually in the process of reconstituting itself, shaping its meanders under the influence of recognizable and desired currents" (6). Anyone following the field can attest that the notion of a mainstream has not only become increasingly problematic; it has also become multiethnic and multinational. As the American automobile industry has permeated pretty much every region of the globe, so the influence of every region of the globe has permeated American fiction. Because the car has such a strong and unique resonance in American culture, however, it functions as one of the fixtures of American identity and thus as a powerful literary trope in the fiction of American writers of both sexes and all colors. In exploring cars and women's fiction, we also get a powerful glimpse of cars and American culture. Certainly, the male writers considered in this study echo much of the fascination and many of the same concerns about cars, but my goal has been to establish women writers' equal share in the excitement of automobility. Focusing on women's cultural and technological acumen provides new insight into the intersection of automotive culture and American literature.

Kathryn Hume has commented on the critical impulse to categorize American fiction by race and gender, noting that, despite the value of such studies, they do have "the unavoidable effect of reinforcing our sense that American fiction consists of isolated interest groups" (2). I am, to some extent, complicit in affirming this isolation, though my intention has been to illustrate that women's fiction does not exist in isolation. Looking at the myriad ways that women writers deploy the automobile reminds us that gender divisions, while real, may be less absolute than some have supposed. Showalter asserts, "To recognize that the tradition of American women's writing is exploding, multi-cultural, contradictory, and dispersed is yet not to abandon the critical effort to piece it together, not into a monument, but into a literary quilt that offers a new map of a changing America, an America whose literature and culture must be replotted and remapped" (175). Any map of twentieth-century America must note the inroads the car has made on our nation and our imagination. Women's powerful yet contested relationship with the automobile stands as a "literary quilt," reflecting the vast diversity of American literary culture. The car drives us together: male and female, white and nonwhite, native-born and immigrant.

The car helped to re-race and re-gender the American landscape, a transformation that women's fiction both produces and reinforces. If to be American is to exercise automobility, then to be an American woman writer is to articulate its place in the life and literature of the nation. Reading American women's fiction through the lens of automobility illuminates the extent to which women on wheels have shaped American culture and American literature.

One • Women on Wheels

1. See, e.g., Armstrong; Banta; Kern; Seltzer; and Tichi.

2. Sherrie A. Inness questions whether access to the car imbued women "with a new sense of agency" or simply transformed them into automobile consumers (48). I would assert that it did both.

3. The car literally drove women to greater equality, as Scharff points out how the suffragists used cars and driving as an integral part of their campaign for the right to vote. See Scharff, *Taking the Wheel,* esp. chap. 5: "Spectacle and Emancipation."

4. McShane cites statistics from Baltimore and New Hampshire, with 4.8 percent of Baltimore car owners in 1911 being women and 4.6 percent of New Hampshire driver's licenses held by women (149–51). Scharff notes that 15 percent of new cars registered in Los Angeles in 1914 were registered to women (25). Because no national statistics appear to exist, one must say that while the number of female owners remained small, those women seem to have had a disproportionate impact on the cultural response to female drivers.

5. One can't help imagining the kind of advertising campaign this could initiate: "Need a sex-change operation? Buy a car instead! It's cheaper and more effective!"

6. In particular, Marx cites Henry James's analysis of American culture in *The American Scene:* "The contrast between the machine and the pastoral ideal dramatizes the great issue of our culture. It is the germ, as James puts it, of the most final of all generalizations about America" (353).

7. Cather, however, seems to have been more drawn to the train than the automobile. Cars figure much less prominently in her work than in Stein's or Wharton's.

8. Both Sherrie A. Inness and Nancy Tillman Romalov note the mixed messages regarding young girls and cars in these series books. Their message, Inness argues, is that "females should be comfortable with technological developments, but need not have an in-depth knowledge of their operation. This precludes them from acquiring the mastery necessary for technological advances" (56). Romalov points out that their mixed genre—adventure, romance, and travel narrative—"is indicative of the larger confusions, motives, and tensions of a society in the process of negotiating the changing dynamics and meaning of modern womanhood" (76). I'm

primarily interested in their presentation of class as mitigating the potentially radical power of the automobile and the positioning of female identity between mind and body.

9. Interestingly, Wharton, the great fan of motoring, apparently didn't drive. Photos show her with a chauffeur, and the permits issued by the French government allowing her to drive to restricted areas during World War I all include her chauffeur, Charles A. Cook (Edith Wharton Collection, YCAL MSS42, box 51, folder 1538).

Two · Modernism

1. Much of the information in these first few pages comes from Michael Berger's very informative piece "The Great White Hope on Wheels." Although I draw heavily on his data, I take the discussion in a different direction, using it to set up an analysis of modern literature, race, and gender. But I am very much in debt to his scholarship on the original incidents.

2. Danica Patrick's fourth-place finish in the 2005 Indianapolis 500—the highest ever by a woman driver—proves that the right car is more important than the "right" gender. Backed by David Letterman's money, Patrick had access to the equipment necessary to be competitive, unlike earlier women drivers such as Janet Guthrie, who were hampered by a lack of resources and thus at a significant technical disadvantage.

3. Tichi does a fascinating job of linking technology to literature, suggesting that "fiction and poetry became recognizable as designed assemblies of component parts" and further claiming that the "machine-age text does not contain *representations* of the machine—it too *is* the machine" (16). I admit to being more interested in how the machine functions and in the dialectic between the "gear-and-girder" world and the notion of flux and upheaval, for it seems to me that the car helps to shape that upheaval.

4. See Lacey 125; and Banta 213.

5. Ford Motor Company also employed women, though in fewer numbers than African-American men. Women worked only on the day shift and by 1914 exceeded 2 percent of the total workforce (Butler 30).

6. It is an interesting side note that Henry Ford finally agreed to settle with the union only after his wife, Clara, threatened to leave him if he did not. Thus, women's influence permeated even the automaker's business decisions (see Sorensen 253). Clara took a similar hard-line stand after the death of Henry's son, Edsel, insisting that he give over the company to his grandson, Henry II, rather than leaving it in the hands of Harry Bennett, the head of Ford's internal security, whose goon squads terrorized workers and beat up on union organizers.

7. Ford also followed a remarkably progressive policy in regards to disability, launching studies to determine which jobs could be performed by disabled workers. Yet here, too, one sees the same impulse to align the human body with a machine; the disabled worker's body serves as a compilation of parts, and the employer's job is to find a way to allow these parts to function as effectively as possible. For more on Ford and disability, see my article "William Faulkner and Henry Ford."

8. I am indebted to Cecelia Tichi's book *Shifting Gears,* in which I first ran across this ad.

9. See, e.g., Howard Lawrence Preston.

10. See esp. Huyssen 47; and Felski 62.

11. Indeed, Faulkner stands as the white American writer who deals most significantly with race, surpassing, I would argue, both Melville and Twain in that regard.

12. I must credit my colleague Alan Price, who generously shared with me his unpublished essay "On the Road with Gertrude Stein, Alice B. Toklas, and 'Aunt Pauline': Their War Service, 1916–1919," thus pointing me toward some very rich material.

13. Alyson Tischler explores Stein's own influence in mass marketing. Her prose often turned up in various advertisements, apparently without protest from her. Despite her reputation as among the most difficult of the modernist writers, she clearly challenges any notion of a division between high and popular culture.

14. I thank my colleague Sandy Spanier for letting me know of the striking portrayal of the car in this text—and for her work in bringing the novel to light and into print.

Three • My Mother the Car?

1. For more on the policing of pregnant women, see Susan Bordo, "Are Mothers Persons?" *Unbearable Weight*, 71–97.

2. Yet the mom car has devastating effects on fathers, also according to the Magliozzis. In determining what should be the second car of a family that has just purchased a minivan, they side with the husband's need for a sporty car, telling his wife, "You are already, by buying a minivan, emasculating the poor man to the extent that it's inhumane" (*Why You*). While clearly tongue-in-cheek, this remark also reaffirms that women are expected to limit themselves to mom cars, while men need cars that do not reflect their reproductive status.

3. This approach is not new. R. L. Lionel refers to a 1927 Oldsmobile ad advising women to use the same acumen in comparison shopping for a car as they do for clothes. In 1932 a Nash advertisement demands, "Why Shouldn't We Shop for Motor Cars Just as We Shop for Clothes?" (qtd. in Lionel 396).

4. I'd like to thank Doreen Fowler for listening to my difficulties in dealing with O'Connor and, through her conversation and insight, helping me to get a grip on some of the complexities of a highly complex writer.

5. As Marvin Magalaner notes of Erdrich's work, "Required for life on and off the reservation, the car is at once as familiar as a hat or as a grocery bag, but, at the same time, invested with a mystique that engenders the awe and respect once reserved for venerated natural spirits" (102).

6. The SUV fails to live up to its reputation as a safe car. Keith Bradsher forcefully argues that, on the contrary, SUVs are less safe than most cars.

Four • Getaway Cars

1. Ramsey, in fact, made the drive nearly every year between 1909 and 1964, with the exception of war years. Clearly, what began as a publicity stunt for the Maxwell-Briscoe Motor Company became a personal passion for her. For more on Ramsey and other early women cross-country drivers, see McConnell.

2. She notes that women, historically, have had less access to travel than men and, citing Eric Leed, that journeys are often conceived as "spermatic" (230) or, according to Mary Gordon, as

"flights *from* women" (231). Given the additional association of women with place, the discourse of travel thus tends to reify patriarchal culture and values.

3. It is not only women whose lives may depend on reliable cars. Gay writer Michael Lane, in *Pink Highways*, begins his saga of journeys by describing what happens when his car breaks down and he finds himself stranded in a town with graffiti proclaiming, "NO FAGS" (9).

4. Kris Lackey, in his study of highway narratives, has suggested that the car can have an antisocial and antifamily impact on the road: "By tremendously exaggerating our mobility and power of sight, so that as we turn our wheel and gaze we scan vast surfaces, it has inflated our sense of individual power, will, and significance. It raises the threshold of novelty for its occupants as it encourages impatience and intolerance" (16). This may hold true for some, but women, who are generally more physically at risk than men, rarely forget their own vulnerability. For them the car provides a degree of protection, reminding them of the vagaries of life on the road rather than promoting intolerance.

5. Yet women trippers rarely appear to resort to the thievery openly celebrated in *On the Road*, in which Sal Paradise gleefully recounts his and Dean's ability to steal gas and food from various filling stations and mom-and-pop stores en route. He justifies such action by launching a tirade against the police who pull them over and fine them after an egregious traffic violation: "It was just like an invitation to steal to take our trip-money away from us. They knew we were broke and had no relatives on the road or to wire to for money. The American police are involved in psychological warfare against those Americans who don't frighten them with imposing papers and threats" (136). Regardless of whether Sal's claim has merit, these men exert a sense of entitlement regarding their rights to the road that is largely unseen in women's road trips.

6. Marilyn Wesley, on the other hand, observes of the woman traveler that she "moves out of her traditional position as object of masculine culture, and her active career controverts the fundamental opposition of masculine mobility in an exterior area to feminine restriction to a domestic space" (xv). Women's road trips begin to erase the boundaries between exterior and domestic space, unlocking the door to that "private room-sized empire."

7. Rosemary Marangoly George, e.g., argues that English women colonialists in India "first achieved the kind of authoritative self associated with the modern female subject" and that, due to "racial privileges," these women established "a coherent, unified bourgeois subjecthood" (6, 61). Caren Kaplan warns that choosing deterritorialization is not the same as having it imposed and cautions "against a form of theoretical tourism on the part of the first world critic, where the margin becomes a linguistic or critical vacation, a new poetics of the exotic" ("Deterritorializations" 191).

8. As Scharff points out in a different piece, "Mobility, Women, and the West," the car was also central to the early Civil Rights movement, particularly the bus boycott in Montgomery, Alabama, beginning in late 1955: "Operating out of forty-three dispatch and forty-two pick-up stations, 325 private automobiles arrived every ten minutes between the hours of 5 and 10 A.M. and 1 and 8 P.M., with hourly pick-ups during the rest of the day" (164).

9. While largely beyond the scope of this chapter, the topic of sex and the automobile has a detailed history. The car has long been associated with both male and female sexuality: gendered as female yet marketed as a phallic symbol. It has also been noted as a site for sexual activity. Given the baggage of sexuality that the car carries, it is interesting to note how little the issue

of sex arises in these texts. Clearly, women's road trips embrace a great deal more than sexual freedom. For discussions of cars and sex, see, e.g., Bayley; David L. Lewis; and Marsh & Collett. On the car as a place for sexual activity, see David L. Lewis; Robert S. and Helen Merrill Lynd, *Middletown;* and Beth L. Bailey.

Five • Mobile Homelessness

1. See Jacobson 9.

2. See Belasco, for a full history of auto camping.

3. As Allan Wallis has pointed out, some of the resistance to mobile homes stems from class issues, as middle-class residents seek to keep such structures out of their neighborhoods. But class, he argues, constitutes only part of the issue; the mobility inherent in such homes often proves profoundly unsettling. See Wallis 22.

4. For more on the development of trailer parks see Hart, Rhodes, & Morgan, esp. 11–15. See also Wallis; and Drury.

5. See, e.g., Rossi, *Without Shelter* and *The Homeless: Opposing Viewpoints,* for discussions of the various figures available on the homeless population.

6. Peter Marin argues, however, that scant resources are dedicated to the plight of homeless men, who constitute by far the largest number of the homeless population.

7. The median home value for the state of Massachusetts in 1990 was $162,800, as compared to the national median, $79,100 (U.S. Census Bureau). The prices for the city of Boston would be much higher.

8. For more on the home as a site of heterosexuality, see Valentine 286–89.

9. While both Lopez and Feinberg associate lesbian sexuality with the motorcycle, other writers link it to cars. In Karin Kallmaker's lesbian romance novel *Car Pool,* e.g., relationships and sexuality are articulated through various forms of automotive metaphoric discourse.

10. This novel is one of the few that has generated critical discussion on the role of the car. Both Jaqui Smyth and Dana Heller explore Adele's fixation with automobiles. Smyth, in particular, notes the tension between the home and the car. She views the car as a "secondary" home (127), while I contend that it *displaces* the home and everything home stands for. But we are much in agreement regarding the relationship between domesticity and the car.

Six • Automotive Citizenship

1. See, e.g., Newitz & Wray for a discussion of how "discourses of class and racial difference tend to bleed into one another" (169).

2. The Petersen Automotive Museum in Los Angeles offered an exhibit of lowriders in 2001. The response from the local Chicano community was overwhelming, as first-time museum-goers flocked to see the show, many of them returning multiple times.

3. Cecelia Lawless argues that the story of the bones in the tar pit "acts as an explanatory analogy of her own family's . . . sense of homelessness, of not belonging in the capital-based market of the United States" (373). But I think that the power Estrella finds in her own bones far overcomes any idea of being reduced to forgotten bones. Rather than highlighting her homelessness, it allows her to feel at home in her body.

4. In an interview Viramontes remarks, "At one point, I wanted Estrella to organize and come out and become a tool of the United Farm Workers. That would have been me coming into it saying that she has got to do this or that, where in reality during that time, the UFW information was not widely disseminated. Whole communities of farmer workers were isolated" (qtd. in Dulfano 659).

Alliance of Automobile Manufacturers. "Economic Contributions." www.autoalliance.org/economic/?PHPSESSID=d530752e729c6dfce918ba6d143bc506.

Allison, Dorothy. *Bastard Out of Carolina.* New York: Plume, 1992.

———. *Skin: Talking about Sex, Class & Literature.* Ithaca, NY: Firebrand Books, 1994.

Alvarez, Julia. *How the García Girls Lost Their Accents.* New York: Plume, 1992.

Anzaldúa, Gloria. *Borderlands / La Frontera: The New Mestiza.* San Francisco: Spinsters / Aunt Lute Books, 1987.

Armstrong, Tim. *Modernism, Technology, and the Body: A Cultural Study.* Cambridge: Cambridge UP, 1988.

Automobile Manufacturers' Association (AMA) News Release. "Automobile Trend Highlights Feminine Influence in Car Design." 1961 press kit. Detroit: Automobile Manufacturers' Association. 6.1–3.

Bailey, Beth L. *From Front Porch to Back Seat: Courtship in Twentieth-Century America.* Baltimore: Johns Hopkins UP, 1988.

Bailey, R. C. "A New Psalm." *Oxford Eagle* July 30, 1925.

Baker, Moira P. "'The Politics of *They*': Dorothy Allison's *Bastard Out of Carolina* as Critique of Class, Gender, and Sexual Ideologies." *The World Is Our Home: Society and Culture in Contemporary Southern Writing.* Ed. Jeffrey J. Folks & Nancy Summers Folks. Lexington: UP of Kentucky, 2000. 117–41.

Balsamo, Anne. *Technologies of the Gendered Body: Reading Cyborg Women.* Durham: Duke UP, 1996.

Banks, Russell. *Continental Drift.* New York: Harper & Row, 1985.

Banta, Martha. *Taylored Lives: Narrative Production in the Age of Taylor, Veblen, and Ford.* Chicago: U of Chicago P, 1993.

Barthes, Roland. "The New Citroën." *Mythologies.* 1957. Trans. Annette Lavers. New York: Farrar, Straus & Giroux, 1972. 88–90.

Baudrillard, Jean. *America.* 1986. Trans. Chris Turner. New York: Verso, 1988.

Bayley, Stephen. *Sex, Drink and Fast Cars: The Creation and Consumption of Images.* London: Faber & Faber, 1986.

Behling, Laura L. "Fisher's Bodies: Automobile Advertisements and the Framing of Modern American Female Identity." *Centennial Review* 41.3 (Fall 1997): 515–28.

——. "'The Woman at the Wheel': Marketing Ideal Womanhood, 1915–1934." *Journal of American Culture* 20.3 (Fall 1997): 13–30.

Belasco, Warren James. *Americans on the Road: From AutoCamp to Motel, 1910–1945.* Cambridge: MIT P, 1979.

Berger, Michael L. *The Devil Wagon in God's Country: The Automobile and Social Change in Rural America, 1893–1929.* Hamden: Archon Books, 1979.

——. "The Great White Hope on Wheels." *The Automobile and American Culture.* 1980. Ed. David L. Lewis & Laurence Goldstein. Ann Arbor: U of Michigan P, 1983. 59–70.

——. "Women Drivers: How a Stereotype Kept Distaff Drivers in Their Place." *Road and Track* (May 1985): 56–60.

Berlant, Lauren. *The Queen of America Goes to Washington City: Essays on Sex and Citizenship.* Durham: Duke UP, 1997.

Bhabha. Homi K. "Introduction: Narrating the Nation." *Nation and Narration.* Ed. Homi K. Bhabha. New York: Routledge, 1990. 1–7.

Blackmer, Corrine E. "Selling Taboo Subjects: The Literary Commerce of Gertrude Stein and Carl Van Vechten." *Marketing Modernisms: Self-Promotion, Canonization, Rereading.* Ed. Kevin J.H. Dettmar & Stephen Watt. Ann Arbor: U of Michigan P, 1996. 221–52.

Blassingame, John W. "Introduction." *Bad Nigger! The National Impact of Jack Johnson.* Al-Tony Gilmore. Port Washington, NY: Kennikat, 1975. 3–7.

Boneseal, Joyce. "Women and the Automobile." *From Horses to Horsepower: One Hundredth Anniversary of the Automobile Industry,* vol. 1: *The First Fifty Years: 1896–1949.* Flint, MI: McVey Marketing, 1996. 21–27.

Boorstin, Daniel J. "Editor's Preface." *The American Automobile: A Brief History.* John B. Rae. Chicago: U of Chicago P, 1965.

Bordo, Susan. *Unbearable Weight: Feminism, Western Culture, and the Body.* Berkeley: U of California P, 1993.

Boyle, Kay. *Process.* Ed. Sandra Spanier. Champaign: U of Illinois P, 2001.

Bradsher, Keith. *High and Mighty: The Dangerous Rise of the SUV.* New York: Public Affairs, 2002.

Brady, Mary Pat. *Extinct Lands, Temporal Geographies: Chicana Literature and the Urgency of Space.* Durham: Duke UP, 2002.

Braidotti, Rosi. *Nomadic Subjects: Embodiment and Sexual Difference in Contemporary Feminist Theory.* New York: Columbia UP, 1994.

Butler, John, & LuAnnette. "Women and the Model T." *The Vintage Ford* 27.4 (July–Aug. 1992): 15–30.

Carlson, Satch. "Life in the Fast Lane." *Autoweek* June 8, 1983, 8.

Casey, Roger. *Textual Vehicles: The Automobile in American Literature.* New York: Garland, 1997.

Castillo, Debra A., & María Socorro Tabuenca Córdoba. *Border Women: Writing from La Frontera.* Minneapolis: U of Minnesota P, 2002.

Chandler, Marilyn R. *Dwelling in the Text: Houses in American Fiction.* Berkeley: U of California P, 1991.

Chénetier, Marc. *Beyond Suspicion: New American Fiction Since 1960.* Trans. Elizabeth A. Houlding. Philadelphia: U of Pennsylvania P, 1996.

Chevrolet Motor Division. "Pretty Soon Every Other Guy Who Walks into Your Showroom Will Be a Woman." Detroit: General Motors Corp., 1986.

Cisneros, Sandra. *Caramelo.* New York: Knopf, 2002.

———. *The House on Mango Street.* New York: Vintage, 1989.

Clarke, Deborah. "William Faulkner and Henry Ford: Cars, Men, Bodies, and History as Bunk." *Faulkner and His Contemporaries.* Ed. Ann Abadie & Donald Kartiganer. Jackson: UP of Mississippi, 2004. 93–112.

Clasby, Nancy T. "'The Life You Save May Be Your Own': Flannery O'Connor as a Visionary Artist." *Studies in Short Fiction* 28.4 (Fall 1991): 509–20.

Clifford, James. "Traveling Cultures." *Cultural Studies.* Ed. Lawrence Grossberg, Cary Nelson, & Paula A. Treichler. New York: Routledge, 1992. 96–116.

Codrescu, Andrei. *Road Scholar: Coast to Coast Late in the Century.* New York: Hyperion, 1993.

Coffey, Frank, & Joseph Layden. *America on Wheels: The First 100 Years: 1896–1996.* Companion volume to the PBS special broadcast. Los Angeles: General Publishing Group, 1996.

Coontz, Stephanie. *The Way We Never Were: American Families and the Nostalgia Trap.* New York: Basic Books, 1992.

Crews, Harry. *Car.* New York: Quill, 1972.

Cross, Gary. *An All-Consuming Century: Why Commercialism Won in Modern America.* New York: Columbia UP, 2000.

Dauphinais, Dean D., & Peter M. Gareffa. *Car Crazy: The Official Motor City High-Octane, Turbocharged, Chrome-Plated, Back Road Book of Car Culture.* Detroit: Visible Ink P, 1996.

DeLillo, Don. *White Noise.* New York: Penguin, 1985.

Deloria, Philip J. *Indians in Unexpected Places.* Lawrence: UP of Kansas, 2004.

Dettelbach, Cynthia Golomb. *In the Driver's Seat: The Automobile in American Literature and Popular Culture.* Westport: Greenwood, 1976.

Didion, Joan. *Play It as It Lays.* 1970. New York: Farrar, Straus & Giroux, 1990.

Dobie, Kathy. *The Only Girl in the Car: A Memoir.* New York: Dial P, 2003.

Domosh, Monu, & Joni Seager. *Putting Women in Place: Feminist Geographers Make Sense of the World.* New York: Guilford P, 2001.

Donofrio, Beverly. *Riding in Cars with Boys: Confessions of a Bad Girl Who Makes Good.* New York: Morrow, 1990.

Dregni, Eric, & Karl Hagstrom Miller. *Ads That Put America on Wheels.* Osceola, WI: Motorbooks International, 1996.

Drury, Margaret J. *Mobile Homes: The Unrecognized Revolution in American Housing.* New York: Praeger, 1972.

Dulfano, Isabel. "Some Thoughts Shared with Helena María Viramontes." *Women's Studies* 30 (2001): 647–62.

Durham, Jimmie. "Geronimo!" *Partial Recall.* Ed. Lucy R. Lippard. New York: New P, 1992. 55–58.

Edgar, J. Clifton, M.D. "The Influence of the Automobile upon Obstetrical and Gynecological Conditions." Report. New York Obstetrical Society, Feb. 14, 1911.

Edith Wharton Collection. Yale Collection of American Literature. Beinecke Rare Book Room and Manuscript Library.

Erdrich, Louise. *The Bingo Palace.* New York: HarperCollins, 1994.

———. *Love Medicine.* New and expanded ed. New York: HarperCollins, 1984, 1993.

———. *Tales of Burning Love.* New York: HarperCollins, 1996.

———. "Where I Ought to Be: A Writer's Sense of Place." *Louise Erdrich's Love Medicine: A Casebook.* Ed. Hertha D. Sweet Wong. New York: Oxford UP, 2000. 43–50.

Eugenides, Jeffrey. *Middlesex.* New York: Farrar, Straus & Giroux, 2002.

Farr, Finis. *Black Champion: The Life and Times of Jack Johnson.* New York: Scribner, 1964.

Farr, Marie T. "Freedom and Control: Automobiles in American Women's Fiction of the 70s and 80s." *Journal of Popular Culture* 29.2 (Fall 1995): 157–69.

Faulkner, William. *Flags in the Dust.* 1929. New York: Random House, 1973.

———. *Go Down, Moses.* 1942. New York: Vintage International, 1990.

———. *Intruder in the Dust.* 1948. New York: Vintage International, 1991.

———. *The Sound and the Fury.* 1929. The Corrected Text. New York: Vintage International, 1990.

Fauset, Jessie Redmon. *Plum Bun.* 1928. Boston: Routledge & Kegan Paul, 1985.

Feinberg, Leslie. *Stone Butch Blues.* Ithaca, NY: Firebrand Books, 1993.

Felski, Rita. *The Gender of Modernity.* Cambridge: Harvard UP, 1995.

Finch, Christopher. *Highways to Heaven: The AUTO Biography of America.* New York: Harper-Collins, 1992.

Fischer, Pam. "Taking the Worries Out of Driving Pregnant." *Going Places* (Mar.–Apr. 2005): 10.

Fitzgerald, F. Scott. *The Great Gatsby.* 1925. New York: Scribner, 1995.

Flink, James J. *The Automobile Age.* Cambridge: MIT P, 1988.

———. *The Car Culture.* Cambridge: MIT P, 1975.

Ford, Henry. In collaboration with Samuel Crowther. *My Life and Work.* New York: Doubleday, Page, 1922.

Ford, Henry. *My Philosophy of Industry.* Authorized interview by Fay Leone Faurote. New York: Coward-McCann, 1929.

Ford, Richard. *Independence Day.* New York: Vintage, 1995.

Ford Motor Co. "Ford: The Universal Car." Dearborn: Ford Motor Co., 1912.

———. "The Woman and the Ford." Dearborn: Ford Motor Co., 1912.

Ford News Bureau. "Design: Women's Influence." ACC 536, box 28. Jan. 10, 1952. From the collection of the Henry Ford Museum and Greenfield Village Research Center.

Frazier, Ian. "No Phone, No Pool, No Pets: Living in a Van." *Atlantic Monthly* Mar. 1996: 48–49.

Fuerbringer, Jonathan. "Junk Ratings Make a Big Splash, Ripples to Follow." *New York Times* May 6, 2005. www.nytimes.com/2005/05/06/automobiles/06bond.html.

Garber, Marjorie. *Sex and Real Estate: Why We Love Houses.* New York: Pantheon, 2000.

Garcia, Cristina. *Dreaming in Cuban.* New York: Ballantine, 1992.

Gardiner, Judith Kegan. "Introduction." *Provoking Agents: Gender and Agency in Theory and Practice.* Ed. Judith Kegan Gardiner. Urbana: U of Illinois P, 1995. 1–20.

Gartman, David. *Auto Opium: A Social History of American Automobile Design.* New York: Routledge, 1994.

Garvey, Ellen Gruber. "Reframing the Bicycle: Advertising-Supported Magazines and Scorching Women." *American Quarterly* 47.1 (Mar. 1995): 66–101.

George, Rosemary Marangoly. *The Politics of Home: Postcolonial Relocations and Twentieth-Century Fiction.* Cambridge: Cambridge UP, 1996.

Gilmore, Leigh. "A Signature of Lesbian Autobiography: 'Gertrice/Altrude.'" *Autobiography and Questions of Gender.* Ed. Shirley Neuman. Portland, OR: Frank Cass, 1991. 56–75.

Gilroy, Paul. "Driving While Black." *Car Cultures.* Ed. Daniel Miller. New York: Berg, 2001. 81–104.

Gramsci, Antonio. *Prison Notebooks, Volume II.* 1975. Ed. and trans. Joseph A. Buttigieg. New York: Columbia UP, 1996.

Haraway, Donna J. "A Cyborg Manifesto: Science, Technology, and Socialist-Feminism in the Late Twentieth Century." *Simians, Cyborgs, and Women: The Reinvention of Nature.* London: Free Association Books, 1991. 149–81.

Harjo, Joy. "The Place of Origins." *Partial Recall.* Ed. Lucy R. Lippard. New York: New P, 1992. 89–93.

Harper, Mary Walker. "The Woman Who Drives a Car Should Know How to Buy and Run One." *Ladies Home Journal* Sept. 1915: 42–43.

Harris, Maxine. *Sisters of the Shadow.* Norman: U of Oklahoma P, 1991.

Hart, John Fraser, Michelle J. Rhodes, & John T. Morgan. *The Unknown World of the Mobile Home.* Baltimore: Johns Hopkins UP, 2002.

Hartigan, John, Jr. "Objectifying 'Poor Whites' and 'White Trash' in Detroit." *White Trash: Race and Class in America.* Ed. Matt Wray & Annalee Newitz. New York: Routledge, 1997. 41–56.

Harvey, David. *The Condition of Postmodernity.* Cambridge: Blackwell, 1990.

Hazleton, Lesley. *Confessions of a Fast Woman.* New York: Addison-Wesley, 1992.

———. *Everything Women Always Wanted to Know about Cars (but Didn't Know Who to Ask).* New York: Doubleday, 1995.

Healy, Patrick O'Gilfoil. "The American Dream, Arriving on Wheels." *New York Times* July 3, 2005. Real Estate sec. 12.

Heller, Dana. A. *The Feminization of Quest-Romance.* Austin: U of Texas P, 1990.

Herlihy, David V. *Bicycle: The History.* New Haven: Yale UP, 2004.

Hitchcock, Mrs. Sherman A. "A Woman's Viewpoint of Motoring." *Motor* 2.1 (1904): 19.

Hollrah, Patrice E. M. *"The Old Lady Trill, the Victory Yell": The Power of Women in Native American Literature.* New York: Routledge, 2004.

Hollreiser, Eric. "Women and Cars." *Adweek's Marketing Week* Feb. 10, 1992: 14–19.

Hu, Pat S., & Timothy R. Reuscher. *Summary of Travel Trends: 2001 National Household Travel Survey.* Federal Highway Administration, Dec. 2004. http://nhts.ornl.gov/2o1/pub/STT.pdf.

Hume, Kathryn. *American Dream, American Nightmare: Fiction since 1960.* Urbana: U of Illinois P, 2000.

"Hummer Owners Pissed over Plan to Make SUV's Safer." *Freepressed* Dec. 8, 2003. www.freepressed.com/suv.htm.

Hurston, Zora Neale. *Jonah's Gourd Vine.* 1934. New York: Harper & Row, 1990.

Huyssen, Andreas. *After the Great Divide: Modernism, Mass Culture, Postmodernism.* Bloomington: Indiana UP, 1986.

Inness, Sherrie A. "On the Road and in the Air: Gender and Technology in Girls' Automobile and Airplane Serials, 1909–1932." *Journal of Popular Culture* 30.2 (Fall 1996): 47–60.

Irigaray, Luce. *This Sex Which Is Not One.* 1977. Trans. Catherine Porter. Ithaca, NY: Cornell UP, 1985.

Jacobson, Kristin. *Domestic Geographies: Neo-Domestic American Fiction.* Diss. Pennsylvania State U, 2004.

Jasper, James M. *Restless Nation: Starting Over in America.* Chicago: U of Chicago P, 2000.

Jerome, John. *The Death of the Automobile.* New York: Norton, 1972.

Joans, Barbara. *Bike Lust: Harleys, Women, and American Society.* Madison: U of Wisconsin P, 2001.

Jones, Kathleen B. "Introduction." *Hypatia* 12.4 (Fall 1997): 1–5.

Joseph, May. *Nomadic Identities: The Performance of Citizenship.* Minneapolis: U of Minnesota P, 1999.

Kadohata, Cynthia. *The Floating World.* New York: Ballantine, 1989.

Kallmaker, Karin. *Car Pool.* Tallahassee: Naiad P, 1993.

Kaplan, Caren. "Deterritorializations: The Rewriting of Home and Exile in Western Feminist Discourse." *Cultural Critique* 6 (Spring 1987): 187–98.

———. "Transporting the Subject: Technologies of Mobility and Location in an Era of Globalization." *PMLA* 117.1 (Jan. 2002): 32–42.

Kennedy, Michelle. *Without a Net: Middle Class and Homeless (with Kids) in America: My Story.* New York: Viking, 2005.

Kennedy, William. *Ironweed.* 1983. New York: Penguin, 1984.

Kern, Steven. *The Culture of Time and Space, 1880–1918.* Cambridge: Harvard UP, 1983.

Kerouac, Jack. *On the Road.* 1957. New York: Penguin, 1991.

Kimmel, Michael. *Manhood in America: A Cultural History.* New York: Free P, 1996.

King, Stephen. *Christine.* New York: Viking, 1983.

Kingsolver, Barbara. *The Bean Trees.* New York: Harper & Row, 1988.

Kingston, Maxine Hong. *China Men.* New York: Knopf, 1980.

Kraig, Beth. "The Liberated Lady Driver." *Midwest Quarterly* 28.3 (Spring 1987): 378–401.

Krugman, Paul. "Toyota, Moving Northward." *New York Times* July 25, 2005. www.nytimes.com/2005/07/25/opinion/25krugman.html.

Kudner Agency, Inc. "For Release Friday, February 1, 1952, and Thereafter." ACC.536, box 1: "Advertising—Wearing apparel, tie-in." From the collections of Henry Ford Museum and Greenfield Village.

Lacey, Robert. *Ford: The Men and the Machine.* Boston: Little, Brown, 1986.

Lackey, Kris. *RoadFrames: The American Highway Narrative.* Lincoln: U of Nebraska P, 1997.

Lane, Michael. *Pink Highways: Tales of Queer Madness on the Open Road.* New York: Birch Lane P, 1995.

Lane, Rose Wilder, & Helen Dore Boyleston. *Travels with Zenobia: Paris to Albania by Model T Ford: A Journal by Rose Wilder Lane and Helen Dore Boylston.* Ed. William Holtz. Columbia: U of Missouri P, 1983.

De Lauretis, Teresa. *Technologies of Gender: Essays on Theory, Film, and Fiction.* Bloomington: Indiana UP, 1987.

Lawless, Cecelia. "Helena María Viramontes' Homing Devices in *Under the Feet of Jesus.*" *Homemaking: Women Writers and the Politics and Poetics of Home.* Ed. Catherine Wiley & Fiona R. Barnes. New York: Garland, 1996. 361–82.

Lawrence, Karen R. *Penelope Voyages: Women and Travel in the British Literary Tradition.* Ithaca, NY: Cornell UP, 1994.

Levitt, Dorothy. *The Woman and the Car: A Chatty Little Handbook for All Women Who Motor or Who Want to Motor.* Ed. C. Byng-Hall. London: John Lane Co., 1909.

Lewis, David L. "Sex and the Automobile: From Rumble Seats to Rockin' Vans." *The Automobile and American Culture.* Ed. David L. Lewis & Laurence Goldstein. Ann Arbor: U of Michigan P, 1980. 123–33.

Lewis, R.W.B. *Edith Wharton: A Biography.* New York: Harper & Row, 1975.

Lewis, Sinclair. *Babbitt.* 1922. New York: New American Library, 1980.

Lionel, R. L. "You've Come a Short Way, Baby." *Automobile Quarterly* 12.4 (1974): 392–99.

Lister, Ruth. "Dialectics of Citizenship." *Hypatia* 12.4 (Fall 1997): 6–26.

London, Jack. "Jack London Says. . . ." *New York Herald* 27 Dec. 1908, 2.3.

Lopez, Erika. *Flaming Iguanas: An Illustrated All-Girl Road Novel Thing.* New York: Simon & Schuster, 1997.

Lynd, Robert S., & Helen Merrell Lynd. *Middletown: A Study in American Culture.* 1929. New York: Harcourt Brace, 1957.

———. *Middletown in Transition: A Study in Cultural Conflicts.* 1937. New York: Harcourt Brace Jovanovich, 1965.

MacComb, Debra Ann. "New Wives for Old: Divorce and the Leisure-Class Marriage Market in Edith Wharton's *The Custom of the Country.*" *American Literature* 68.4 (Dec. 1996): 765–97.

Magalaner, Marvin. "Louise Erdrich: Of Cars, Time, and the River." *American Women Writing Fiction: Memory, Identity, Family, Space.* Ed. Mickey Pearlman. Lexington: UP of Kentucky, 1989. 95–108.

Magliozzi, Tom & Ray. "Dear Tom and Ray." *Car Talk.* www.cartalk.com/content/columns/Archive/1995/June/23.html.

———. *Maternal Combustion: Calls about Moms and Cars.* Compact disc. Cambridge: Dewey, Cheetam & Howe, 2004.

———. *Why You Should Never Listen to Your Father When It Comes to Cars.* Compact disc. Cambridge: Dewey, Cheetam & Howe, 1999.

Mann, Patricia S. "Cyborgean Motherhood and Abortion." *Provoking Agents: Gender and Agency in Theory and Practice.* Ed. Judith Kegan Gardiner. Urbana: U of Illinois P, 1995. 133–51.

Marchand, Roland. *Advertising the American Dream: Making Way for Modernity, 1920–1940.* Berkeley: U of California P, 1985.

Marin, Peter. "Homelessness Mostly Affects Single Men." *The Homeless: Opposing Viewpoints.* Ed. Tamara L. Roleff. San Diego: Greenhaven, 1996. 46–51.

Marks, Patricia. *Bicycles, Bangs, and Bloomers: The New Woman in the Popular Press.* Lexington: UP of Kentucky, 1990.

Marsh, Peter, & Peter Collette. *Driving Passion: The Psychology of the Car.* Boston: Faber & Faber, 1986.

Martin, Biddy, & Chandra Talpade Mohanty. "Feminist Politics: What's Home Got to Do with It?" *Feminist Studies/Critical Studies.* Ed. Teresa de Lauretis. Bloomington: Indiana UP, 1986. 191–212.

Martinez, Manuel Luis. "Telling the Difference between the Border and the Borderlands: Materiality and Theoretical Practice." *Globalization on the Line: Culture, Capital, and Citizenship at U.S. Borders.* Ed. Claudia Sadowski-Smith. New York: Palgrave, 2002. 53–68.

Marx, Leo. *The Machine in the Garden: Technology and the Pastoral Ideal in America.* New York: Oxford UP, 1964.

Mason, Bobbie Ann. *In Country.* New York: Harper & Row, 1985.

Mattera, Joanne. "Women and Cars: On a Roll." *Glamour* May 1990: 254–59.

McAlister, Melani. "Can This Nation Be Saved?" *American Literary History* 15.2 (Summer 2003): 422–40.

McConnell, Curt. *"A Reliable Car and a Woman Who Knows It": The First Coast-to-Coast Auto Trips by Women, 1899–1916.* Jefferson, NC: McFarland & Co., 2000.

McDowell, Linda. *Gender, Identity and Place: Understanding Feminist Geographies.* Minneapolis: U of Minnesota P, 1999.

McShane, Clay. *Down the Asphalt Path: The Automobile and the American City.* New York: Columbia UP, 1994.

Meese, Elizabeth A. *(Sem)erotics: Theorizing Lesbian: Writing.* New York: New York UP, 1992.

Messer-Davidow, Ellen. "Acting Otherwise." *Provoking Agents: Gender and Agency in Theory and Practice.* Ed. Judith Kegan Gardiner. Urbana: U of Illinois P, 1995. 23–51.

Michael, Magali Cornier. "Re-imagining Agency: Toni Morrison's *Paradise.*" *African American Review* 36.4 (Winter 2002): 643–61.

Morris, Mary. "Women and Journeys: Inner and Outer." *Temperamental Journeys: Essays on the Modern Literature of Travel.* Ed. Michael Kowalewski. Athens: U of Georgia P, 1992. 25–32.

Morris, Meaghan. "At Henry Parkes Motel." *Cultural Studies* 2.1 (Jan. 1988): 1–47.

Morrison, Toni. *Paradise.* New York: Penguin, 1999.

———. *Song of Solomon.* 1977. New York: Penguin, 1987.

Moya, Paula M. L. *Learning from Experience: Minority Identities, Multicultural Struggles.* Berkeley: U of California P, 2002.

Mumford, Lewis. *The Highway and the City.* New York: Harcourt, Brace & World, 1963.

National Automobile Dealers Association. "Auto Retail Industry Strong, Resilient in 2001, NADA Data Shows." www.nada.org/Content/NavigationMenu/Newsroom/News_Releases/2002/ind_05_08_02.htm.

Newitz, Annalee, & Matthew Wray. "What Is 'White Trash'? Stereotypes and Economic Conditions of Poor Whites in the United States." *Whiteness: A Critical Reader.* Ed. Mike Hill. New York: New York UP, 1997. 168–84.

Oates, Joyce Carol. "Lover." *Granta 58: Ambition* (June 1997): 159–69.

———. *them.* New York: Fawcett Crest, 1969.

O'Connor, Flannery. "A Good Man Is Hard to Find." *A Good Man Is Hard to Find and Other Stories.* New York: Harcourt, Brace & Co., 1955. 9–29.

———. "The Life You Save May Be Your Own." *A Good Man Is Hard to Find and Other Stories.* New York: Harcourt, Brace & Co., 1955. 53–68.

———. *Wise Blood.* New York: Farrar, Straus & Giroux, 1952.

Owens, Louis. *Mixedblood Messages: Literature, Film, Family, Place.* Norman: U of Oklahoma P, 1998.

Patterson, Martha H. "Incorporating the New Woman in Wharton's *The Custom of the Country.*" *Studies in American Fiction* 62.2 (Fall 1998): 213–36.

Penrose, Margaret. *The Motor Girls in the Mountains.* New York: Cupples & Leon Co., 1917.

———. *The Motor Girls on a Tour.* New York: Cupples & Leon Co., 1910.

Petry, Ann. *The Street.* Boston: Houghton Mifflin, 1946.

Piercy, Marge. *The Longings of Women.* New York: Fawcett, 1994.

Pierson, Michelle Holbrook. *The Perfect Vehicle: What It Is about Motorcycles.* New York: Norton, 1997.

Pirsig, Robert M. *Zen and the Art of Motorcycle Maintenance.* New York: Morrow, 1974.

Porter, Richard C. *Economics at the Wheel: The Costs of Cars and Drivers.* San Diego: Academic P, 1999.

Powers, Lyall H., ed. *Henry James and Edith Wharton Letters: 1900–1915.* New York: Scribner, 1990.

Pratt, Mary Louise. *Imperial Eyes: Travel Writing and Transculturation.* New York: Routledge, 1992.

Preston, Claire. *Edith Wharton's Social Register.* New York: St. Martin's, 2000.

Preston, Howard Lawrence. *Dirt Roads to Dixie: Accessibility and Modernism in the South, 1885–1935.* Knoxville: U of Tennessee P, 1991.

Price, Alan. "On the Road with Gertrude Stein, Alice B. Toklas, and 'Aunt Pauline': Their War Service, 1916–1919." MS. 1–34.

Primeau, Ronald. *The Romance of the Road: The Literature of the American Highway.* Bowling Green, OH: Bowling Green State U Popular P, 1996.

Pritchard, Elizabeth A. "The Way Out West: Development and the Rhetoric of Mobility in Postmodern Feminist Theory." *Hypatia* 15.3 (Summer 2000): 45–72.

Rae, John B. *The American Automobile: A Brief History.* Chicago: U of Chicago P, 1965.

Ralston, Hugh C. "Memo to Charles E. Carll." Aug. 9, 1951. ACC 536, box 45. From the collection of the Henry Ford Museum and Greenfield Village Research Center.

Ramsey, Alice Huyler. *Veil, Duster, and Tire Iron.* Pasadena: Castle P, 1961.

Reck, Franklin M. *A Car Traveling People: How the Automobile Has Changed the Life of Americans—A Study of Social Effects.* Detroit: Automobile Manufacturers Association, 1945.

Rich, Adrienne. "Notes toward a Politics of Location (1984)." *Blood, Bread, and Poetry.* New York: Norton, 1986. 210–31.

Roberts, Randy. *Papa Jack: Jack Johnson and the Era of White Hopes.* New York: Free P, 1983.

Robinson, Marilynne. *Housekeeping.* 1981. New York: Bantam, 1982.

Rollins, Montgomery. "Sane Motoring—or Insane?" *Outlook* 29 (May 1909): 275–82.

Romalov, Nancy Tillman. "Mobile and Modern Heroines: Early Twentieth-Century Girls' Automobile Series." *Nancy Drew and Company: Culture, Gender, and Girls' Series.* Ed. Sherrie A. Inness. Bowling Green, OH: Bowling Green State U Popular P, 1997. 75–88.

Ross, Kristin. *Fast Cars, Clean Bodies: Decolonization and the Reordering of French Culture.* Cambridge: MIT P, 1995.

Ross, Marlin B. "Trespassing the Colorline: Aggressive Mobility and Sexual Transgression in the Construction of New Negro Modernity." *Modernism, Inc.: Body, Memory, Capital.* Ed. Jani Scandura & Michael Thurston. New York: New York UP, 2001. 48–67.

Rossi, Peter. *Without Shelter: Homelessness in the 1980s.* New York: Priority P, 1989.

Rothschild, Emma. *Paradise Lost: The Decline of the Auto-Industrial Age.* New York: Random House, 1973.

Roy, Denise. *My Monastery Is a Minivan: Where the Daily Is Divine and the Routine Becomes Prayer: 35 Stories from a Real Life.* Chicago: Loyola P, 2001.

Russell, Betty G. *Silent Sisters: A Study of Homeless Women.* New York: Hemisphere Publishing, 1991.

Rydell, Robert W. *World of Fairs: The Century-of-Progress Expositions.* Chicago: U of Chicago P, 1993.

Saldívar-Hull, Sonia. *Feminism on the Border: Chicana Gender Politics and Literature.* Berkeley: U of California P, 2000.

Scanlon, Jennifer. "Under Whose Direction? Consumer Culture's Message Makers." *The Gender and Consumer Culture Reader.* Ed. Jennifer Scanlon. New York: New York UP, 2000. 195–200.

Scharff, Virginia. "Mobility, Women, and the West." *Over the Edge: Remapping the American West.* Ed. Valerie J. Matsumoto & Blake Allmendinger. Berkeley: U of California P, 1999. 160–71.

———. *Taking the Wheel: Women and the Coming of the Motor Age.* New York: Free P, 1991.

Schneider, Greg. "Automakers Put More Women at the Wheel." *Washington Post* May 1, 2005. http://washingtonpost.com/wp-dyn/content/article/2005/05/01/AR2005050100961.html.

Seltzer, Mark. *Bodies and Machines.* New York: Routledge, 1992.

Senna, Danzy. *Caucasia.* New York: Riverhead Books, 1998.

Shea, Anne. "'Don't Let Them Make You Feel You Did a Crime': Immigration Law, Labor Rights, and Farmworker Testimony." *MELUS* 28.1 (Spring 2003): 123–44.

Showalter, Elaine. *Sister's Choice: Tradition and Change in American Women's Writing.* Oxford: Clarendon P, 1991.

Silko, Leslie Marmon. *Ceremony.* New York: Penguin, 1977.

Simpson, Mona. *Anywhere but Here.* 1986. New York: Vintage, 1992.

Singer, Bayla. "Automobiles and Femininity." *Research in Philosophy and Technology* 13 (1993): 31–41.

Smiley, Jane. *A Thousand Acres.* New York: Fawcett, 1991.

Smith, Sidonie. *Moving Lives: Twentieth-Century Women's Travel Writing.* Minneapolis: U of Minnesota P, 2001.

Smyth, Jacqui. "Getaway Cars and Broken Homes: Searching for the American Dream *Anywhere but Here.*" *Frontiers* 20.2 (1999): 115–32.

Sorensen, Charles E. With Samuel T. Williamson. *My Forty Years with Ford.* New York: Collier, 1956.

Spain, Daphne. *Gendered Spaces.* Chapel Hill: U of North Carolina P, 1992.

Spanier, Sandra. "Introduction." *Process.* Kay Boyle. Urbana: U of Illinois P, 2001.

Stanley, Alessandra. "Do Women Have the Zoom Gene?" *New York Times* Apr. 26, 2002. D1+.

Stein, Gertrude. *The Autobiography of Alice B. Toklas.* New York: Random House, 1933.

Steinbeck, John. *Cannery Row.* New York: Viking, 1945.

———. *The Grapes of Wrath.* 1939. New York: Penguin, 1992.

Steinem, Gloria. "Sex, Lies & Advertising." *Ms.* July–Aug. 1990: 18–28.

Stern, Jane, & Michael. *Auto Ads.* New York: Random House, 1978.

St. James, Lyn. "Women Tell Automakers Not to Sell Them Short." *Detroit Free Press* Mar. 22, 1993: 1E.

Stokes, Katherine. *The Motor Maids across the Continent.* New York: Hurst & Co., 1911.

Stout, Janis P. *Through the Window, Out the Door: Women's Narratives of Departure from Austin and Cather to Tyler, Morrison, and Didion.* Tuscaloosa: U of Alabama P, 1998.

"SUVs Capture New Female Buyers." www.roadandtravel.com/newsworthy/Newsworthy2002/womensuvs.htm.

Taylor, Janelle S. "The Public Fetus and the Family Car: From Abortion Politics to a Volvo Advertisement." *Public Culture* 4.2 (Spring 1992): 67–80.

Thomas, Brook. "*China Men, United States v. Wong Kim Ark,* and the Question of Citizenship." *American Quarterly* 50.4 (1998): 689–717.

Thomas, Kelly L. "White Trash Lesbianism: Dorothy Allison's Queer Politics." *Gender Reconstructions: Pornography and Perversions in Literature and Culture.* Ed. Cindy L. Carlson, Robert L. Mazzola, & Susan M. Bernardo. Burlington, VT: Ashgate, 2002. 167–88.

Tichi, Cecelia. *High Lonesome: The American Culture of Country Music.* Chapel Hill: U of North Carolina P, 1994.

———. *Shifting Gears: Technology, Literature, Culture in Modernist America.* Chapel Hill: U of North Carolina P, 1987.

Tischler, Alyson. "A Rose Is a Pose: Steinian Modernism and Mass Culture." *Journal of Modern Literature* 26.3–4 (Spring 2003): 12–27.

Toklas, Alice B. *The Alice B. Toklas Cook Book.* New York: Harper & Brothers, 1954.

Tompkins, Jane. *Sensational Designs: The Cultural Work of American Fiction, 1790–1860.* New York: Oxford UP, 1985.

Toomer, Jean. *Cane.* 1923. New York: Liveright, 1975.

Trask, Michael. *Cruising Modernism: Class and Sexuality in American Literature and Social Thought.* Ithaca, NY: Cornell UP, 2003.

Tuttle, Cameron. *The Bad Girl's Guide to the Open Road.* San Francisco: Chronicle Books, 1999.

Twitchell, James B. *Adcult USA: The Triumph of Advertising in American Culture.* New York: Columbia UP, 1996.

U.S. Census Bureau. "Historical Census of Housing Tables: Home Values." Feb. 18, 2005. www.census.gov/hhes/www/housing/census/historic/values.html.

Valentine, Gill. "(Hetero)Sexing Space: Lesbian Perceptions and Experiences of Everyday Spaces." *Space, Gender, Knowledge: Feminist Readings.* Ed. Linda McDowell & Joanne P. Sharp. New York: John Wiley & Sons, 1997. 284–300.

Viramontes, Helena María. *Under the Feet of Jesus.* New York: Penguin, 1995.

Vogel, Paula. *How I Learned to Drive.* New York: Dramatists Play Service, 1997.

Wald, Matthew L., & David D. Kirkpatrick. "Congress May Require Closer Scrutiny to Get a Driver's License." *New York Times* May 3, 2005. www.nytimes.com/2005/05/03/politics/03licenses.html.

Wallis, Allan D. *Wheel Estate: The Rise and Decline of Mobile Homes.* New York: Oxford UP, 1991.

Watkins, Evan. "'For the Time Being, Forever': Social Position and the Art of Automobile Maintenance." *Boundary 2* 18.2 (Summer 1991): 150–65.

Weiss, Ruth. "Auto Companies Heed Advice from Women on Color, Styling." *Detroit News* Sept. 16, 1956: 1F.

Wesley, Marilyn C. *Secret Journeys: The Trope of Women's Travel in American Literature.* Albany: SUNY P, 1999.

Wharton, Edith. *The Custom of the Country.* 1913. New York: Scribner, 1956.

———. *The House of Mirth.* 1905. New York: New American Library, 1964.

———. *A Motor-Flight Through France.* New York: Charles Scribner's Sons, 1908.

"Why Women Are, or Are Not, Good Chauffeuses." *Outing Magazine* (May 1904): 154–59.

Wicke, Jennifer. *Advertising Fictions: Literature, Advertisement, and Social Reading.* New York: Columbia UP, 1988.

Wolfe, Tom. *The Kandy-Kolored Tangerine-Flake Streamline Baby.* New York: Farrar, Straus & Giroux, 1965.

Wolff, Janet. "On the Road Again: Metaphors of Travel in Cultural Criticism." *Cultural Studies* 7.2 (May 1993): 224–39.

Woolf, Virginia. *Three Guineas.* 1938. New York: Harcourt Brace Jovanovich, 1966.

Wright, Priscilla Hovey. *The Car Belongs to Mother.* With Illustrations by Carl Rose. Boston: Houghton Mifflin, 1939.

Wright, Richard. *Native Son.* 1940. New York: Harper & Row, 1966.

Young, Andi. "Selling Cars to Women: A Changing Market." *Automotive Age* Oct. 1982: 10–16.

Young, Clarence. *The Motor Boys Overland.* New York: Cupples & Leon, 1906.

Page numbers followed by *f* indicate illustrations.